Building and Detailing
MODEL AIRCRAFT

PAT HAWKEY

Kalmbach Books
21027 Crossroads Circle
Waukesha, Wisconsin 53186
www.Kalmbach.com/Books

© 2009 Pat Hawkey.
All rights reserved. This book may not be reproduced in part or in whole by any means whether electronic or otherwise without written permission of the publisher except for brief excerpts for review.

Published in 2009
13 12 11 10 09   1 2 3 4 5

Manufactured in the United States of America

ISBN: 978-0-89024-723-5

Front cover (also pages 11 and 38-39, top): 1/48 scale Tamiya P-51B, photos © Kalmbach Books

Publisher's Cataloging-In-Publication Data

Hawkey, Pat.

  Building and detailing model aircraft / Pat Hawkey.
    p. : ill. (chiefly col.) ; cm. – (FineScale modeler books)

  At head of title on cover: Scale modeler's how-to guide
  ISBN: 978-0-89024-723-5

  1. Airplanes--Models--Handbooks, manuals, etc. 2. Airplanes--Models--Design and construction--Handbooks, manuals, etc. 3. Airplanes--Models--Finishing--Handbooks, manuals, etc. 4. Models and modelmaking. I. Title. II. Title: Model aircraft III. Title: Scale modeler's how-to-guide

TL770 .H39 2009
629.133/134

Introduction 4

1 Where it all began 6

2 Game plan 11

3 Paint, tools, and glue 16

4 Building almost "out of the box" 23
1/48 scale SBD Dauntless

5 Enhancing a modern "shake-and-bake" kit 38
1/48 scale P-51B Mustang

6 Scribing surface details..................... 47
1/48 scale RF-84 Thunderflash

7 Adding realism with aftermarket details 56

   Applying photoetched metal 60
   1/32 scale Fi.156 Storch

   Combining kits and adding resin parts..... 66
   1/48 scale Spitfire Mk.IXe

   Handling vacuum-formed canopies 75
   1/72 scale P-40N Warhawk

8 Vacuum-formed and resin conversions 82
1/72 scale F-106B Delta Dart, TF-102 Delta Dagger

About the author 95

Introduction

Creating miniature replicas must be something hard wired into the brains of a certain percentage of us humans. Evidence of this activity can be found nearly as far back as there are human records. For those wishing to express themselves creatively in this way, or those simply wanting to own a representation of a favorite flying machine, things have never been better.

The plastic kit industry and aftermarket accessory manufacturers are producing the highest-quality items the modeling world has ever seen. If you're just discovering this world of styrene, your timing is superb. If you're returning from an absence, prepare to be impressed. You are benefiting from the result of years of serious modelers making their thoughts and wishes known to those who produce the goods.

I hope anyone who is a fan of model airplanes will be entertained by what's between these covers and will be encouraged to advance his or her skills. It would be helpful to know the basic modeling techniques before attempting the projects demonstrated here. Readers I think will benefit most probably are already using an airbrush. Not that airbrushing is all that hard; it's just that investing in painting equipment and spending the time to learn our individual set-ups and paint preferences form a kind of threshold in the hobby we all have had to cross before enjoying the greatest rewards.

My goal is to help you build on your skills with what you learn in this book so that you'll have the confidence of knowing just about any desirable modeling goal is within reach. Many of these old modeling tricks were a normal part of kit construction before the advent of today's well-detailed and tight-fitting models. So this book may also give you an appreciation for the humble beginnings of the hobby. I wasn't there at the dawn of plastic models, but I had a glimpse of the dark ages.

My qualifications for writing this book are unique. Like most young boys in the '60s, I grew up on plastic model kits. Unlike my peers, I never gave it up. I certainly noticed cars and girls in my teen years, but a real love of airplanes and creating accurate little plastic ones were powerful forces in me too. A huge influence was the proximity of the original Squadron Shop hobby shop in Hazel Park, Michigan. Besides the books, tools, decals, and exotic imported kits, there were showcases full of built models.

This was a time of *Scale Modeler* magazine in the U.S.; *Scale Models* and *Airfix* magazines from England. Period. No Internet to surf over endless postings of model images. Half an hour away on my bike though, I could study what for all I knew was the cutting edge in built models. What I saw in those cases pointed to where I wanted to go. I honed my skills until my models shared shelf space with

Its silver and gloss-white paint scheme "makes" this Academy 1/72 scale C-97. However, a wonderful paint job only works when the foundation—the model—has been well made and when its surface has been properly prepared to reveal such a smooth finish.

the others at the Squad Shop. By the late '70s, mine were being noticed too, and I actually sold a few. (Who knew what to charge? In fact, I practically gave them away. What counted to me was that I had been paid for the product of something I did for fun. That was heady stuff.)

A few years later, through some fortuitous contacts, I found myself building models for other people on a regular basis. In 1986, I created Hawx Planes and officially turned my hobby into a business. It just seemed like the thing to do. From then until now, I've never had to wait by the phone for someone to order a model airplane. Between models and writing articles for *FineScale Modeler* magazine, I've always been busy.

So what you have in this book is one professional's approach to building and detailing plastic model airplanes. My way of building may differ from that of many hobbyists because of my concern with speed and efficiency. I don't mean to imply that I run a production line, but it is true that the faster I can get a high-quality piece out the door, the faster I get paid and the sooner I can move on to the next model. I can't afford to lavish hour upon hour on a model if it doesn't substantially improve the results. I do not have the luxury of taking five steps to do something that I can do with reasonable results in two. (I don't believe that making things complicated necessarily makes them good.)

If I absorbed anything from my days hanging around the Squadron Shop it was the importance of accuracy. Above all, accuracy means getting the aircraft shape right, then being diligent about markings, finish, weapons configurations, and the correct angles of open canopies, doors, and flying surfaces. Searching for and studying pictures of the real thing can easily take as much time as building and finishing the model. For me, it's still an essential part of the process, and you'll see in this book how it influences some of the things I do.

One of the most difficult aspects of this book has been choosing what subjects best illustrate what I think is important in building and detailing. I've tried to include enough variety in kit quality, subject, and scale to keep things interesting for everybody.

Notice the key words in the title of this book are building and detailing—no mention of "superdetailing." Where does detailing stop and superdetailing begin? I don't know. (I do know that the less work I have to do with tweezers, the happier I am.) I hope that with what you see in this book, you'll master some detailing basics and go on to develop your detailing skills as far as you want to take them. Maybe you will write the superdetailing book!

1
Where it all began

During World War II, the first "plastic" (Bakelite or cellulose acetate) model airplanes called ID models were produced for military recognition training. Best known of these 1/72 scale ID model producers was the Frog Company of England. Another was the Cruver Company of Chicago. These simple models were, for the most part, one-piece solid objects intended only to provide aircraft silhouettes that aircrews and ground observes would likely see. Rarely, if ever, would they have been painted or decorated in any way.

Almost 40 years of incremental kit improvement can be seen in this comparison of OOB (out-of-the-box) 1/72 scale Bristol Beaufighters. The Airfix kit (top) was released in the late 1950s or early 1960s. The Hasegawa counterpart arrived on the scene in 1998. Besides the apparent outline differences, there's raised rivet detail on the Airfix kit vs. recessed panel lines on the Hasegawa and much sharper and more-accurate detail with the Japanese release. The Airfix kit isn't in the same league as the Hasegawa version for fit, especially in the clear parts department. At the time of this writing, the Airfix kit is still being produced, unchanged from its glory days decades ago.

Anyone remember these? From the early '60s, these were strictly entertainment for kids. Markings were printed on thin paper and were to be cut out and glued on. "Hobby Miniatures" and "Made in Hong Kong" are the only clues as to where they came from. Scale—who knows?

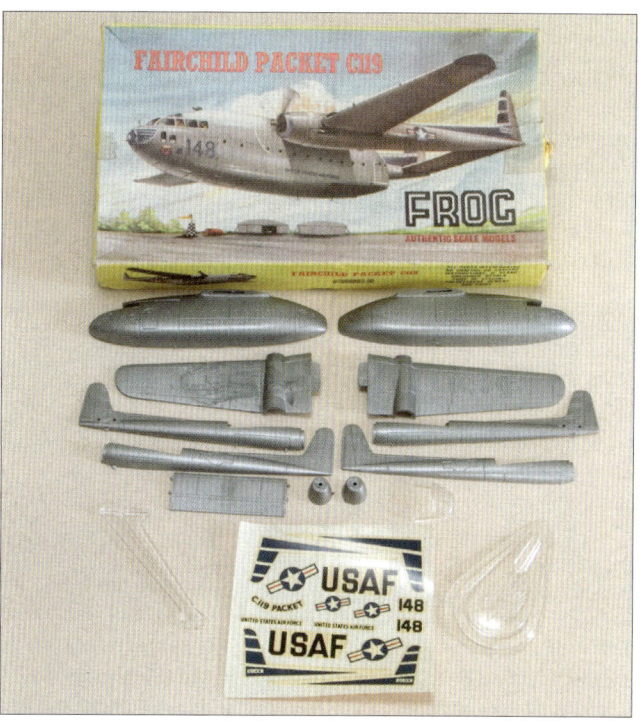

Here's an interesting choice of subject, also from the early '60s. Keeping things simple was important to manufacturers in those days. A display stand used to be standard issue in a model airplane kit—with or without landing gear. Locations for decals are molded in. This one "scales out" at 1/179.

After the war, Frog (named from the company's slogan, "Flies Right Off the Ground," referring to the wood flying models the company originally produced) began creating plastic multi-part model airplanes. Since Frog's plastic models did not fly, this new line of product was christened Penguin. At first they were odd-scale, simple models with few parts and no detail. They were, in fact, only slightly more complicated wartime ID models. They were successful though, and eventually Frog re-adopted its standard 1/72nd scale from the ID models. The number of kits grew and began to include landing gear and other "details."

A competitor to Frog, Airfix, appeared in the early '50s also producing its kits in 1/72 scale. While making a line of models in the same scale (so-called constant scale) seems like an obvious idea today, in the early years of plastic model airplanes, scale was anything but constant. Manufacturers were more inclined to make models fit the size of their models' packaging, resulting in what were called box-scale models. (A B-52 made to fit in a 12"-long box will be in a much different scale than a P-51 made to fit that same box.)

An excellent example of a model from the early years of plastic airplanes is the Revell F-106. As soon as (if not sooner than) the prototype was rolled out, molds were being cut of the most exciting subject of the day. Box art and decals accurately represent one of the pre-production Delta Darts. The kit is a mix of accurate outline, moving parts, and remarkable recessed surface detail. Unfortunately part of the nice recessed detail is those darned decal location marks. Cockpit detail is a seat and a pilot. Missing is the classic ball-and-swivel stand Revell included in its early releases. Scale is close to 1/72.

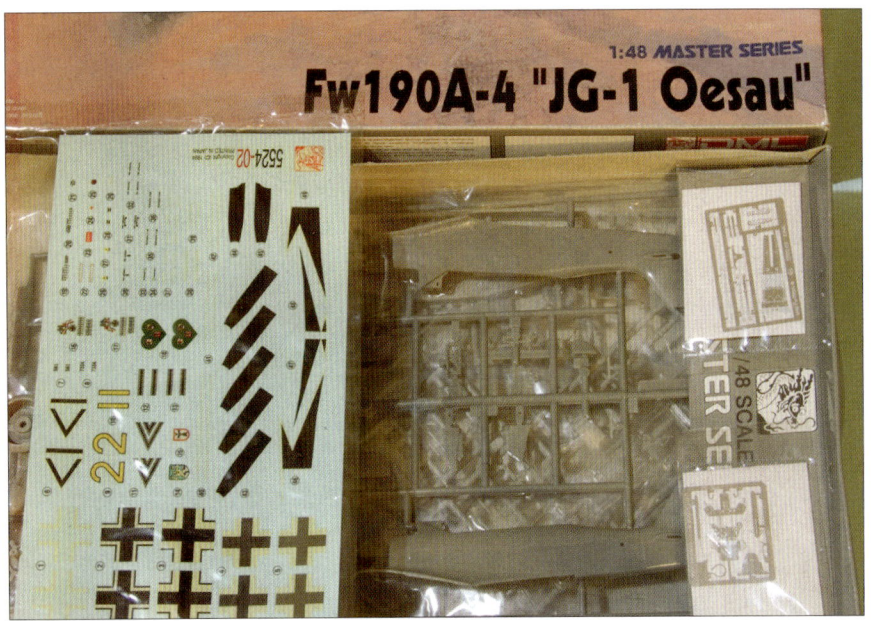

DML released this ex-Tri-Master kit in 1994 as part of its Master Series. Tri-Master's white metal parts had been replaced with plastic, but the photo-etched metal (PE) frets were still part of the package. Of interest is what this master kit does not include. No swastika appears either on the box art or on the decal sheet. Display of the printed swastika is illegal in parts of Europe.

Up from toys. In the United States, injection-molded plastic kits began to appear in significant numbers in the 1950s. Companies including Hawk, Comet, ITC (Ideal Toy Company), Revell, and Lindberg produced the very first. As did Frog in England, some of the first kit companies offered both wood and plastic lines of models. At the time, it seems, "real" modelers and craftsmen worked in wood—plastic kits were for kids.

Thus, plastic models of the day were simplistic, often with no landing gear. Cockpit "detail" was likely just a pilot's head, and sometimes clear parts were not part of the package. (Paint the window areas blue, the instructions would suggest.) As often as not, decal locations were molded right into the plastic—eventually to become so disliked by serious modelers when they needed to be filled and sanded that they were referred to as "idiot marks."

Despite the industry's seemingly jumbled beginnings, kids loved building models, and the plastic model airplane hobby was here to stay. Soon now-familiar companies like Monogram, Aurora, and Renwal joined the party.

The late '50s and early '60s was a time of dramatic growth and excitement in the real airplane world. Jet aircraft were finally getting the powerful engines they needed, and prototypes of all descriptions were being introduced in large numbers. Kit manufacturers had no shortage of subjects. Unfortunately, with a general blanket of security covering things new and military in those Cold War days, accuracy was far from assured. On the other hand, what did the target audience—young males—know or care about accuracy? Kit sales were good.

Japanese invasion. We're used to thinking of Japan as a leader in production of world-class automobile and electronic products, so it's easy to forget that most of us first saw the stamp "Made in Japan" on cheap plastic toys and other inexpensive consumer products. To many, it was a derisive term. However, it wasn't much of a leap from toys to plastic model kits, and it wasn't long before Japan set the quality standards for accuracy and precision in models as well as in cars and TVs. Although it's not surprising that Japanese companies got into the plastic kit business, what is amazing is the large number of model manufacturers that small nation produced.

Like kit makers elsewhere, the original Japanese offerings were aimed at youngsters as much as anyone. Scales were all over the map, but in general were notably smaller than U.S. and European kits. Although 1/144 scale was becoming accepted for airliners, certain Japanese companies introduced many prop and jet fighter types in this small scale. For a short time, 1/70 and 1/50 scales caught on in Japan—but only there. Companies like Hasegawa and Tamiya that would one day be setting the industry standards got started in this period.

In the U.S., kits quickly became more sophisticated from an engineering standpoint, but they were still thought of as toys. As a result, moving parts and what were called "action features" became selling points. Lindberg in particular appealed to young electricians by including electric motor kits to assemble with the plastic parts in order to "motorize" the model.

Standards arrive. In 1963, Revell introduced the hugely popular 1/72 scale WWII "Warbird" line, followed closely by WWI kits for those fans of the smaller scale. Monogram, Lindberg and Renwal added 1/72 scale subjects to the market, but Revell was the big contributor in the U.S. For those who liked their models bigger, 1/48th scale was becoming something of a standard as well. Monogram started to introduce its line of single-engine WWII classics to complement Hawk's line in this scale. Perhaps as a sign that manufacturers were beginning to understand the maturity of serious modelers, "idiot marks" became a thing of the past.

1963 was also a very significant year in plastic modeling for a different reason. A group of builders in England met to create the British Plastic Modeler's Society (BPMS). Within a year, there was an American branch and the "British" in the title changed to "International." IPMS was to grow quickly and become very much an international organization.

The modeling world expanded to a record size in the mid to late '60s. A war in Vietnam kept military matters in constant focus. Major war films like *Patton*, *The Blue Max*, *Tora, Tora, Tora*, and *The Battle of Britain* also fueled interest. Kids and adults were building plastic model kits, and manufacturers responded. British model makers could be counted on for a couple of new releases every month.

IPMS membership grew, and the organization provided an umbrella under

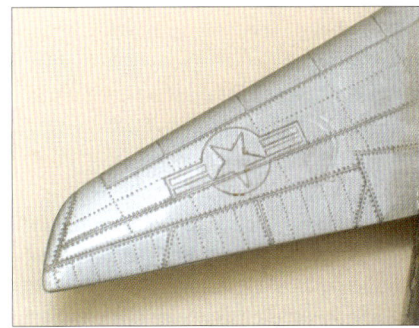

Here's a great example of a decal location molded into the model, in this case, on a Comet F-100 Super Sabre (also later boxed by Aurora). In this case, compounding the misery of dealing with what modelers came to call "idiot marks," is that the mark is upside down.

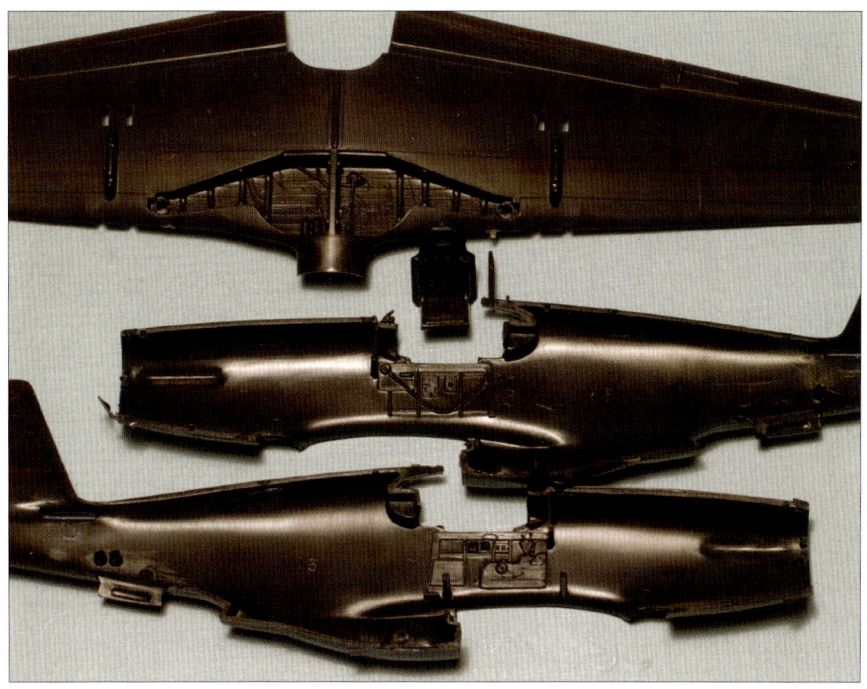

Monogram's P-51B was something of a breakthrough kit. Instead of working parts (other than the standard spinning propeller), you received an amazingly detailed interior, seat, and wheel wells. Those wheel wells were something else in 1967!

which adults and youngsters could gather to share experience. IPMS gave the model building public a voice it hadn't had before as well as some influence on the kits that manufacturers chose to produce.

Plastic airplane models weren't just for kids anymore: Moving parts with "play value" began to disappear, and more-realistic detail and scale accuracy replaced gimmicks. Aircraft features previously ignored such as wheel wells and cockpits received more and more attention. Odd scales were largely history. The industry began to settle on 1/144, 1/72, and 1/48 scales, and in 1967, Revell added 1/32 scale with the introduction of a P-40, a Spitfire, and a Bf109 in that scale. Hasegawa soon followed with its own offerings, and this big scale became hugely popular.

During the '60s, kits began to be imported and distributed by companies other than those that produced the original molds. A company called UPC packaged and distributed kits from at least 13 different kit producers. AMT boxed Hasegawa and Frog kits. Frog boxed Hasegawa kits. MPC boxed Airfix kits. Air Lines boxed Frog kits. In almost all cases, the company doing the new packaging also printed the decals. Speaking from personal experience, those decals were almost always useless, but the kits could be found anyplace toys were sold.

Seventies slow-down. The '70s brought the first slow-down of the plastic kit industry. Styrene, being a petroleum product, suffered from the reduced oil supply from the Middle East. For this and other reasons, some of the long-established names in the plastic kit industry, including Aurora, Hawk, and Frog, folded. Many of their molds would eventually be obtained by other companies, and those kits would be seen again in new packaging. Those companies that stayed in the game cut back on new tooling. U.S. companies focused on reissues of older kits.

The Japanese company Otaki got the model airplane world's attention with 1/48 scale renditions of classic WWII fighters featuring exquisite recessed panel lines. Hasegawa and Fujimi were doing the same with select 1/72 scale subjects. Besides just looking so much better than raised lines and rivets, recessed lines made the assembly—especially the seam filling and sanding—significantly easier. If recessed lines disappeared under putty, you just reclaimed them with your favorite scribing tool—there's no easy way to rebuild lost raised panel lines. U.S. and European kit manufacturers were slow to pick up on this important feature. The fervent wish of many builders, myself included, was for kits with Monogram-type interior detail and Otaki-type external detail.

An aftermarket evolves. While new kit production slowed, resin and rubber mold materials and technology advanced. Enterprising folks realized they could produce limited runs of specialized plastic hobby items at a fraction of the cost of the major kit producers, who were committed to expensive steel injection molds appropriate for tens of thousands of plastic "shots."

Enthusiasts in their basements and garages carved wood masters, poured rubber over them, filled the molds with resin, and sold the results. Thus was born what would eventually be known as the resin aftermarket industry. Besides detail and conversion parts, entire airplane kits were produced. And thanks to these "garage businesses," the hobby didn't lack for new items or excitement in times when the big-name companies weren't cranking out new kits the way they had in the good old days.

In 1988, a small company in Japan called Tri-Master brought out a limited number of 1/48 scale Luftwaffe fighter kits. Squadron's catalog described them as "museum quality, high tech," which meant a mixed-media kit of a styrene model with some photo-etched and white metal parts included. These were aimed squarely at serious builders and collectors.

Even by today's standards, they were expensive, but these kits made an impact. "High-tech" and "high-grade" kits from

A KIT COMPARISON

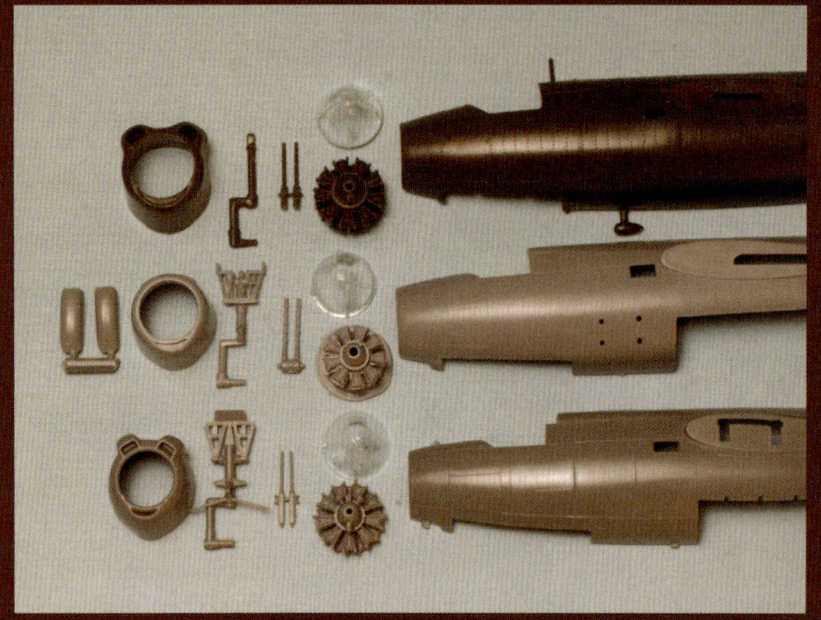

There is no doubt that in 40 years on the market, model kits have improved, revealing more detail and greater accuracy. On top is Revell's late-'60s release of the 1/72 scale B-26 Marauder; in the middle is Airfix's take on the subject from 1974; on the bottom is Hasegawa's offering of 2007. Revell's simplistic kit is covered with raised rivets. Note the main gear strut compared to the other two. The shape of the cowling is questionable and has some mighty sink marks on the upper scoops. The kit consists of slightly less than 60 parts and originally came with a price of $1.50. Airfix raised the bar with its rendition. Though the surface detail is raised, the totally inaccurate rivets are gone. This kit featured detailed wheel wells as well as two full bomb bays, plus bombs. Landing gear and engines more closely resemble the real things, but there was a step backward with the defensive armament. Two markings options and clear landing light lenses and wingtip navigation lights were nice touches. With about 150 parts, the kit cost $3.50 when first released. Then Hasegawa showed how it's supposed to be done. Recessed surface detail, extremely accurate shapes, and superfine detail inside and out are on display. What the image doesn't convey well is the improvement in the clear parts over the previous two kits. (This is especially important to bombers with all their windows and turrets.) There are about 170 parts and three markings options. Suggested price at its release was a little over $40.

other manufacturers soon began to appear. These were simply existing kits repackaged to include a small fret of photo-etched items or white metal or resin replacement parts. Such kits may or may not have been improved with additional non-styrene pieces, but they were made trickier to build.

Serving the serious modeler. In 1994, the American company Accurate Miniatures hit the scene and raised quality standards a little higher with 1/48 scale Allison Mustangs. Like Otaki in the '70s, Accurate picked a popular subject that had not previously been well represented, and what a job the company did! For the first time, it was obvious a kit had been engineered and produced by a crew that understood what was important to experienced modelers.

Here was the fantastic recessed external detail combined with great interior-molding that modelers had wished for. This was combined with intelligent informative instructions, and a very nice decal sheet. AM also chose a subject that could be kitted in a number of slightly different versions. Since the Allison-powered Mustangs weren't nearly as well known and popular as the B, C, and D versions, Accurate Miniatures seemed to have its sights on the more serious builders and collectors.

In 2001, China established itself firmly in the plastic kit world with releases by Trumpeter. The company wasn't timid, starting with big 1/32 scale kits of Shenyangs (Chinese MiGs), followed by an A-10. Within two years, the company was producing 14 1/32 scale kits, and by 2004, Trumpeter picked up where Airfix had left off in the early 1970s with 1/24 scale WWII types as well.

These physically larger models have proven to be popular, despite their cost. In 2009, the average age of modelers was over 50 according to *FineScale Modeler* magazine surveys. Vision isn't all that it once was for those of us who welcomed diminutive 1/144 and 1/72 scale kits back in the '70s. Still, anyone who's looked at a big Trumpeter kit will have noticed many, many little pieces. Good eyes or bad, small parts are still a fact of life in the hobby. After all, it's about miniaturizing things.

Plenty for all. The result of this industry's history is that there are about 50 years' worth of plastic airplane kits to choose from. For the beginner, a reasonable rule of thumb is that the newer the release, the higher the quality and accuracy. But one must be aware that a new release doesn't necessarily mean a new plastic kit. The steel molds the kits are born from are expensive to produce and have an extremely long lifespan. A cold, unused mold isn't generating revenue. So don't be surprised that long-established kit companies re-release old kits from time to time in contemporary packaging.

While the companies of the 1960s that simply printed boxes and decals to market kits made by other manufacturers are gone, some kit companies have arrangements with other companies to supplement their own "Y" product lines with the help of some of the other's "Z" molds.

What this means for the modeler is that the much-hyped, all-new kit may not be made from the latest research used to make accurate new molds. Rarely, if ever, is the lineage of a such a repackaged product mentioned on the "Y" box. Let the buyer beware—and let the wise buyer ask fellow modelers at local club meetings or on the favorite on-line forum about particular kits before purchase.

2

Game plan

Fancy cockpits and open engine compartments and gunbays are of secondary importance to me. Job one is to build a smooth, realistically symmetrical airframe that will eventually wear a convincing paint and decal finish.

Detail is delightful, but it doesn't compensate for an ugly step at the wingroot or mismatched dihedral (a wing's upward angle relative to horizontal), a crooked landing gear, or a canopy that looks like it was an afterthought. If the first impression a model gives is "I'm crooked," how can you take it seriously? Why would you look harder? Attention to the basics will invite closer scrutiny, whereas that detail stuff will make a good model a great model.

A good approach to building the overall model is more important than the number or type of detail parts added. This Tamiya P-51 is enhanced with additional details as shown in Chapter 5.

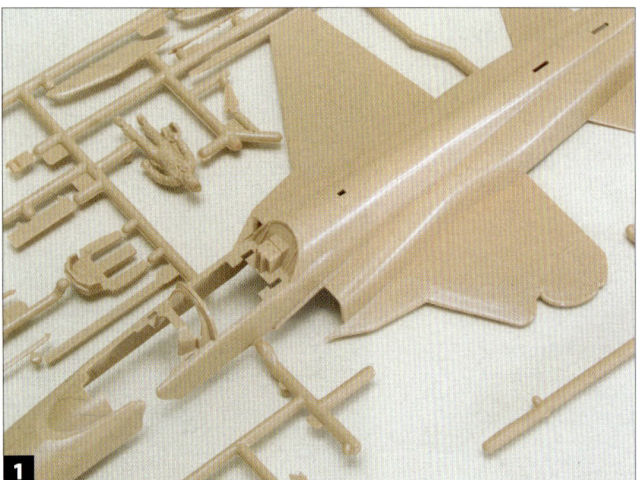

1 Sometimes things go amiss at the model factory and the molten styrene doesn't completely fill the mold cavity of a sprue. This is called a short shot and is illustrated here. Note the missing left wingtip and the gap in the nose. Remedy? Send the kit back to where you got it.

2 Single-edge razor blades are extremely handy in separating model parts from their sprues. These blades are thinner than hobby-knife blades, lessening the chances of damaging parts when removing them from sprues.

3 Specialized nippers or manicure scissors like those shown here do a clean job of removing parts from sprues. If you're going to be sanding the piece anyway, twisting off the part works too.

4 Here's a good way to protect clear parts while the model is being built: Bag 'em and pin the bag someplace it will easily be found later on.

Parts preparation. I don't think I've ever read a how-to-build-models book that didn't suggest that the first step should be to wash the parts in warm soapy water. This is done in order to remove any of the oily mold release agent that may be clinging to the styrene from contact with the mold that produced it. The reasoning is that paint and/or glue will not adhere to the plastic if this film is in the way.

This is probably good advice, and I recommend you follow it until you've developed your own working habits. For the record, I've never, ever bathed my model parts before assembly. Shame on me, I guess, but I know why it's not necessary given the way I build. In my case, any light oil film isn't a factor in the gluing process and by time any of my models reach the serious painting stage, they will have been well bathed.

Once the box has been opened and the contents have been checked for warped wings, twisted fuselages, or short shots (where the plastic hasn't filled the mold cavity completely), **1**, I generally detach the big pieces from the sprues and start dry-fitting. Another time-honored admonition is to never twist parts to break them from the sprues. Except for small detail parts, I do it all the time. For me, the time I spend touching up a little plastic nib—especially in an area I will be sanding and filling anyway—is more than offset by not having to reach for another tool. For those small detail parts, a single-edged razor blade used like a guillotine will remove them cleanly from their location on the sprue. (Buy yourself a 100-count box of these cheap wonder cutters, **2**.) Manicure scissors, **3**, are good for this as well. And you can always get that dedicated sprue nipper at your hobby shop.

The above suggestion, so contrary to traditional advise, is not a vote for carelessness. Rather it's an encouragement for you to think through each step and develop a workflow that's right for you.

The case for clear parts. Clear parts are always a special case. Most kit instructions

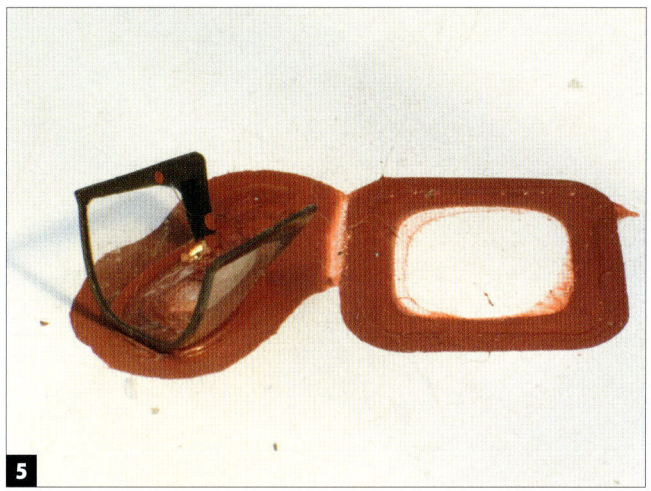

5 Murphy's Law is always at work at a modeler's workbench, even the bench of a modeler putting together a book on the subject. Here's a mishap that was not staged for the camera. I'll let your imagination fill in the blanks. Protect your clear parts!

6 An important step in preparing clear parts is to dip them in Future floor polish. Use a container big enough to accommodate the clear part. Hold it by an edge or a corner that will be painted, dip the part, and allow the excess to drip off.

7 Set the part on paper towel to wick away the remaining excess Future and place a cover over it to protect it from dust while the acrylic coating dries.

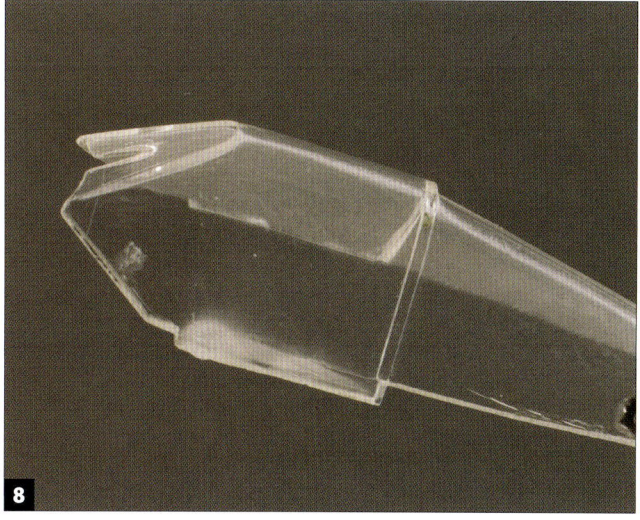

8 Here's what super glue can do to unprotected clear styrene. You don't want to see the white frost that forms around the edges from super glue creeping up the sides of your canopy as it cures. You can polish it away if you can get at it, but avoiding it with a protective layer of Future is the way to go.

call for the canopy to be the last thing glued to the model. Don't let that suggest to you that you don't worry about it until the very end of the project. I give the clear parts priority at an early stage. I am fussy about clear parts on a model, believing they can make or break the look of the finished piece. Finding out they don't fit at the end of the story, when options are few, is bad news indeed.

Almost any other detail part can be replaced by a bit of scratchbuilding or sifting through the spares box, but a canopy that gets broken or goes missing in action will stop the project in a hurry. Once I determine that the kit-supplied glass is usable, I usually set it aside for protection by inserting it in a clear plastic envelope and pinning it to a bulletin board until needed, **4**, **5**. If you put the clear parts in some other safe place, don't forget where it is. Oh yeah, I've lost them that way.

One procedure that has become standard for me is giving the clear parts a dip in Future acrylic floor finish. This not only gives parts a glistening finish that looks great on the model, it provides a barrier to the fumes from super glue that otherwise can cause the clear plastic to frost, **8**. It's critical to get the piece as free of dust as possible before dunking it. A good shot of compressed air should accomplish this. Then, hold the part in tweezers and submerge it in a container of Future. Remove it and allow the excess to drip off the part, **6**. Touching the dripping part to the side of the container expedites this step. Place the part edges down on a piece of paper towel to wick away excess Future, **7**, and find some kind of dust cover for the part while the Future dries. With luck, in a couple of hours, you'll find a shiny clear part without any dust or cat hair imbedded in the clear coating. If you do find blemishes, scrub the part lightly in ammo-

9 The blue halo around this bomb trapeze is the most obvious of parting lines. Called "flash," this is usually the result of styrene being forced out of a worn mold. Sometimes this stuff is worth collecting for use as thin shimming material.

10 Parting lines are usually present on leading edges of single-piece stabilizers or wings. Even if they are subtle, be assured a coat of paint will make the lines look worse. They are easily removed with a few passes of a fine-grit sanding stick or a Flex-I-File.

11 To remove a parting line from a rounded surface like most tires, a Flex-I-File is the tool of choice. It's a strip of sanding material on a harp or bow, which allows the strip to conform to the object.

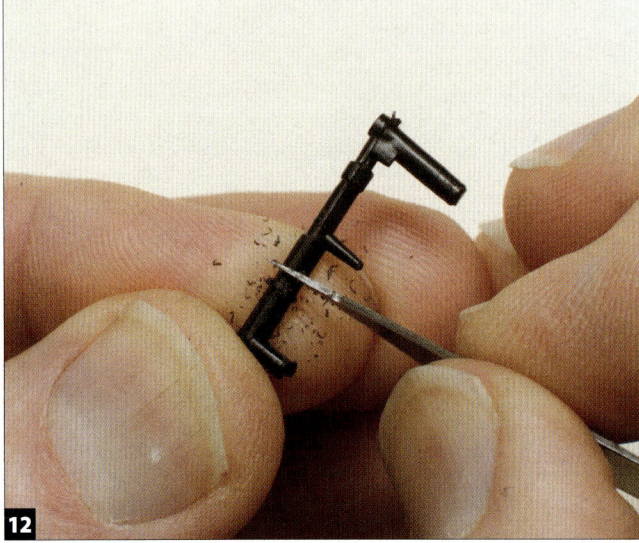

12 Scraping with a fresh knife blade is a good method for cleaning the mold lines off of detail parts.

nia to remove all the Future. It won't harm the styrene in the least, and you're set to try it again.

Be aware that Future makes a part shiny, not clear. It won't transform a yellowed opaque plastic piece into a beautiful clear one.

Dash the flash. Because the injection-molding process consists of two molds pressed together while molten styrene is injected into them, every single piece in the kit will have a parting line. This marks where the two halves of the mold came together around the given piece. If the mold is worn, styrene may escape through a parting line to form flash, **9**. On a flat piece, say a stabilizer, the parting lines will be around all the edges and may not be too noticeable.

Noticeable or not, the lines will be there, and a few swipes with a medium-to-fine sanding stick should make them go away, **10**. Another good tool for this is a Flex-I-File. This is a ribbon of abrasive material stretched between aluminum arms. Pressing it against a part causes it to conform well to curved surfaces, **11**.

On fine, round parts like landing gear struts and gun barrels, the lines will be pretty obvious. To remove these lines, scrape with a fresh No. 11 blade. Remember that your goal is just to remove the raised line. You don't want to scrape a flat surface onto the round one.

I routinely remove molded-on antenna masts, pitot tubes, gun barrels, and raised lights early on, **13**. Because they're molded squarely on seam lines, it's inconvenient to work around this stuff. It's much more efficient in the long run to remove these detail parts, deal with seams and the rest of the construction unhindered, and then return them or replace them at the end of the job. Replacing is an espe-

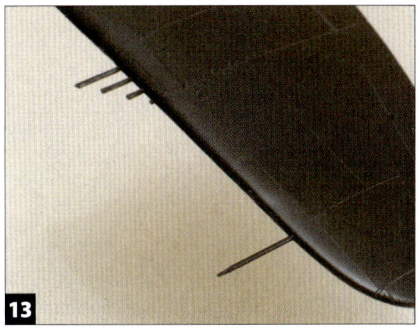
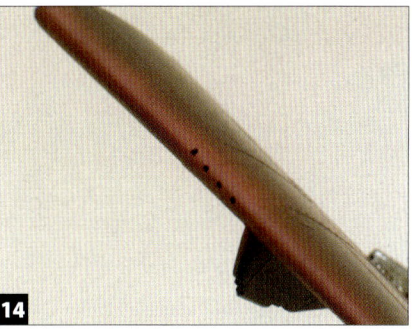

13 Although the kit manufacturers feel obligated to include all the detail parts, sometimes the way they do it isn't builder friendly. It would be silly to work around these fragile guns and pitot tube while building this model. The kit-supplied items work best as templates for your own (better) replacement parts.

14 The wing halves of this P-47 have been glued together, the guns and pitot tube have been removed, and holes have been drilled for styrene rod replacement parts. In the case of the guns, the holes are drilled in more-accurate positions than the kit locations.

15 With the kit parts serving as templates, styrene rod guns and a rod-and-sprue pitot tube have been substituted for the molded detail.

16 Just like car tires, aircraft tires are somewhat flat on the patch that touches the ground. A couple swipes on an emery board or medium-grit sandpaper will put just enough "flat" in a plastic tire for your purposes.

17 You'll know you have it right when the wheels stand straight up by themselves. Sand off just enough material from the tires to make them do this.

cially good idea with pitot tubes and gun barrels or anything else round in cross-section. You can probably make a more convincing detail piece from plastic or metal tube and/or rod material than what is supplied in the kit, **14**, **15**.

Inside out. Most fighter-type kits are engineered so that the internal things like cockpit tubs and tail wheels are in place before the fuselage goes together. If at all possible, I like to get the fuselage together first, making sure seams mate, are strong, and the fuselage has a good fit with wings and other major components. This may mean removing or modifying pins, ledges, or other locating devices inside the fuselage. By separating the interior from the exterior for as long as possible, I can work on one while the other is drying.

A standard modification I apply to the tires of any airplane that will be parked is to sand a flat spot on each. I simply grab the wheel firmly between thumb and forefinger and scrape it at a 90-degree angle across an emery board until the wheel sits straight up, **16**. At first, the wheel is likely to lean slightly one way or another, but a couple more swipes with a sanding stick will level the spot. You'll know you've got it right when the wheel stands 90 degrees to the ground by itself, **17**.

Of course you can buy resin wheels that are pre-flattened and bulged for that "weighted look." Some contemporary kits even offer them as an option along with full-round wheels. The theory is correct, but to my eyes, weighted tires usually look deflated. If I must use bulged tires, I sand the sides of the bulges down some.

In the excitement to get building on a brand-new model, it's easy to overlook the steps in this game-planning stage. However, looking things over carefully, preparing parts, and having a plan will save hours in fix-ups later on.

So, ready to build a model airplane?

3
Paint, tools, and glue

A solid paint job is essential if your goal is creating a realistic model.

Shouldn't that be tools, glue, then paint? Maybe, but in most instances, painting interiors has to be accomplished before serious assembly takes place. Kit manufacturers pay a lot of attention to internal areas—cockpits in particular—these days, and perfectly adequate detail may reveal itself when you open your kit box. With some simple painting techniques, you'll be amazed at the look you can achieve without resorting to aftermarket upgrades. So let's talk about paint first.

A bit about paint. If you plan to take your model building to anything approaching a level you'd want to show your friends, you'll want to get an airbrush and learn to use it. There's no way around that one. With practice, a person can become proficient with an ordinary paintbrush, but if you like soft edges between colors and subtle weathering effects, you need an airbrush.

You don't need to spend a fortune to begin with, and I strongly suggest you don't. You can always upgrade your airbrush later, and you'll probably find continued uses for your original equipment. All the models you see in this book were

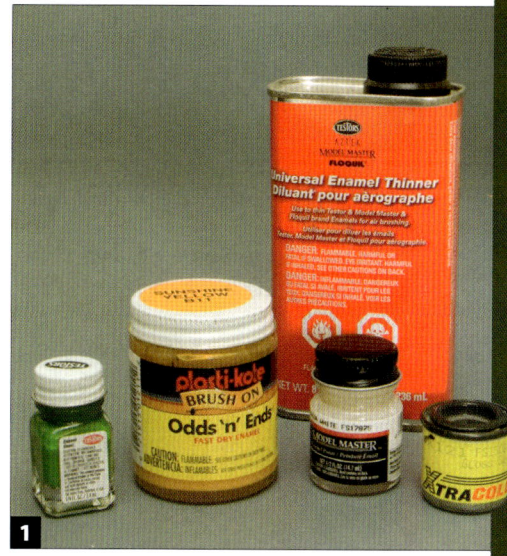

Here's an assortment of gloss enamel brands I use. From left to right are Testors classic, Plasti-Kote, Testors Model Master, and a tin of Extracolor. Plasti-Kote can be found in hardware stores, is of good quality, and dries fast. Consider it if you are comfortable mixing your own colors. If not, let Extracolor do it for you. The thinner is Testors Universal Enamel Thinner, which will cut all the enamels seen here.

airbrushed exclusively with a basic single-action Paasche H. That's the tool I learned to use more than 40 years ago, and it's still my front-line piece of equipment. (For more on airbrushing, see *How to Use an Airbush, 2nd Edition* (2008) from Kalmbach Books.)

Even though the book you're reading now is about building and detailing model aircraft, it's hard to describe how to create a beautifully finished model without some mention of painting techniques. Probably nothing in modeling is more personalized than how you paint. What follows, then, is a description of my methods, some of them admittedly old school, some no doubt superseded by new technology—but they work. Put the same tools and materials that I describe here in the hands of another modeler, and you'll get a different result. I hope you can use these few paragraphs to help flatten out the learning curve for you.

I ordinarily don't prime any portion of a model other than the areas I've worked over with sandpaper. I prime these areas both to reveal and to fill scratches. Likely as not, the primer will be thoroughly sanded away before any paint goes on. These days I decant spray can auto primer (I'm not fussy about brand names) and airbrush that. If I want a thin mix, I thin it with ordinary lacquer thinner. Since I don't use a lot of primer, a spray can lasts a long time.

Get a respirator. This whole model airplane building activity involves many chemicals and materials unfriendly to humans. One really does need to be aware and protect one's self accordingly. Take the warnings on the labels seriously. Ventilate your work area. (And as long as we're talking about safety, take extra care with knives and saws for hobby work. When you start putting pressure on something with a sharp edge, be aware of where it's likely to go if it slips. And when your hobby knife starts rolling across the bench, it's important to move your feet—ask me how I know that one.)

I grew up using enamels, Testor and Pactra mostly, and enamels are what I understand and continue to use, **1**.

Much more environmentally and people-friendly are acrylic paints. They've come a long way from their introduction in the early '80s. I use acrylics primarily for touch-up and detail brush painting, because they won't interact with the enamels underneath them the way another enamel will, and acrylics can certainly be airbrushed. As a general rule though, acrylics don't seem to be as forgiving of mixing different brands the way enamels are. If you choose acrylics, it's a good idea to find a line of paint that works well for you and stay with it as well as that line's particular thinner. In a bind, a "universal" thinner that has worked well for me when I have airbrushed acrylics is ordinary automotive window-washing solvent.

Skip the flats. I almost never airbrush flat paints. Gloss enamels airbrush much smoother. They also provide the smooth surface that decals need to cling to. If starting with gloss paint, there is no need for that coat of clear gloss near the end of the project. That's a step eliminated and time saved.

How about gloss for camouflage colors? I don't seem to have a problem with mixing my own. With the limited palette available when I was a kid, I was forced to

2

The first step in creating depth with a "sludge wash" is to slop a dark gray acrylic wash over a base coat of gloss paint. In this case the color is a Chromate Green.

3

When the wash dries, rub a damp cotton swab over the area, picking up all the dark gray except that in the corners, nooks, and crannies of the raised detail.

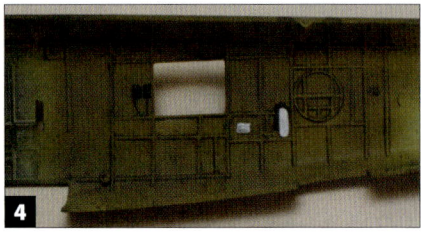

4

With the excess wash removed, apply a clear flat coat—you'll see some of the detail start to be picked out.

5

After painting the various bits of molded-in detail, further highlight the raised areas with a light dry-brushing of silver. That's it. Now take a picture, because once you put the fuselage together, you won't see it again.

come up with some of my own shades. It's not rocket science. If you don't have exactly the color you want or need, don't be afraid to experiment. White with a touch of black and a touch of yellow is Gloss Gull Gray. Hit it with flat coat and you've got flat Gull Gray. Brown plus green makes olive. Black plus yellow produces great shades of chromate green or yellow depending on the ratios.

If the idea of mixing your own paint is scary, there's the Extracolor line of gloss enamels from England. These are handy from time to time, although I don't always agree with Extracolor's interpretation of some colors. (But then, there are people who don't agree with my interpretation of some color.)

You will often see "FS numbers" for colors. These numbers come from the Federal Standard 595, which was established by the United States government during World War II to standardize paint colors on equipment as it left the production line. There was a time when a simple written request could get you a copy of this document and all the color chips for free. Today, it'll run you a couple hundred dollars. However, you can go online to see the colors in question. Type "federal standard 595" into your search engine and a num-

ber of sites will come up. At these, you can track down that five-digit number and see exactly what that color the instructions call for looks like, provided you have a relatively accurate color monitor.

A dirty wash. The most basic painting detail technique is the wash. This is simply a thin dark layer of paint put on over a base coat to fill the low spots, nooks, and crannies. In effect, it's forcing scale shadows into areas that don't have enough relief for light to do the job by itself. The easiest way I know to do this is the method christened the "sludge wash" by long-time *FineScale Modeler* senior editor Paul Boyer.

The wash itself is simply acrylic paint thinned at least 50 percent with water and about 30 percent plain dish soap added, **2**. The soap breaks the surface tension of the water, allowing it to flow easily into every recess, and in larger amounts also does a number on the adhesive qualities of the paint on a smooth gloss surface. With the proper amount of soap in the mix, the dry paint will rub off the surface with a damp cloth, **3**. The paint protected from the cloth in the recessed areas will stay put, **4**. This technique will create your shadows and greatly enhance the three-dimensional quality of the molded-in detail, **5**.

I start by putting the base color coat down in gloss enamel. It is essential this be gloss, not flat, otherwise the wash will only result in a big permanent stain. Once the base is dry, a wash of dark gray goes over the area. I only use true black on silver areas (like landing gear and wheels), as it gives a stark appearance anyplace else. I don't actually measure the paint-soap-water ratios. Usually I dip my brush in dish soap and dab that onto a palette of some sort. (I like the backs of surplus photos or

retired CDs.) Then I add about triple the amount of water to the soap and a dab of the paint I'm going to use. I alter the mix until it looks and acts "right." You want more soap in the mix than it takes to just break the water tension, but you don't want so much that the mix doesn't want to flow. Experiment with your mix until you get the feel of what it's going to do.

Pre-shading and dry-brushing. Another effective way to achieve the look of shadows in the recesses is to pre-shade all the ribs and stringers and areas around and behind pieces of internal equipment with a significantly darker shade of whatever color you're using in that area, **6**. This is a bit more time consuming and requires a steady hand and fine-line qualities with your airbrush. Once the dark lines are "drawn," the true interior color can be applied, but not heavily enough that the darker areas disappear. The idea is to have the dark show through the true color coat, **7**. This pre-shading works better as the scale of the model increases. It's impractical on anything smaller than 1/48th.

To get fine raised details, such as those in cockpit interiors, to pop out, dry-brushing gives the best results, **8**. The term dry-brushing is a bit misleading. Damp brushing would be a more accurate description of what's going on. You want enough paint in the bristles of your brush to leave pigment behind when you press them against raised detail but not enough paint to flow. With a little practice, you'll find out how much paint and how much brush pressure will give you the result you desire. I like Testors Model Master Chrome Silver to brush over dark gray or black areas. It stays "alive" on the brush a long time. A couple swipes over any raised areas instantly show you where the raised detail is. You may be happy to leave the

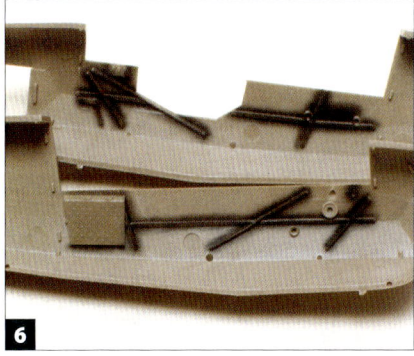

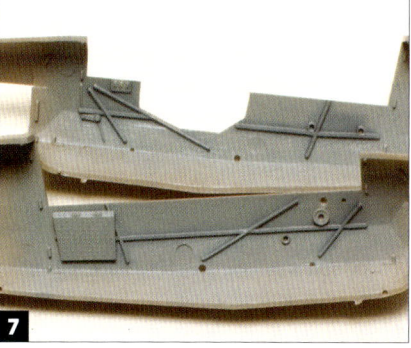

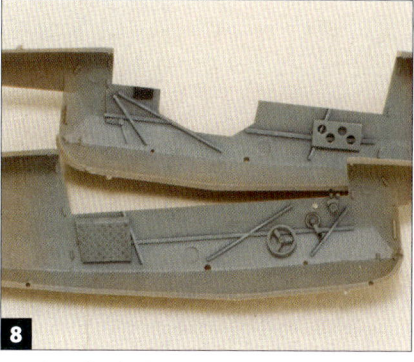

6 To pre-shade an interior, start by painting the raised detail a dark color—in this case, gloss black.

7 Apply the base color (RLM 02 in this case), but not so heavily that it cancels the previous black.

8 Dry-brush a lighter shade of the base coat on the raised detail. Follow with a light sludge wash to detail parts, then spray on a flat clear coat. On a contest model, the ejector-pin marks would be filled in!

raised detail in silver. If you want to pick out any knobs or buttons with something other than silver, they will now be much easier to see.

If you're shooting for an accurate model, be aware that cockpit interiors are not very colorful places. Many modelers overdo this area so that the cockpit ends up looking like a pinball machine, with all kinds of colored buttons and knobs. This is a fantasy look. Pilots know their cockpits by feel. Imagine—you wouldn't rely on a color-coded cockpit if you lost your "electricals" at night. The boring fact is that most knobs are black and toggle switches are silver and there are few colored lights flashing. Warning lights only show color if they are on. If they are on, the engine's running and there had better be a pilot strapped in.

Glue school. The material I first learned to put model pieces together with was plastic cement. Found in tubes of gel or fast-running liquid in bottles, plastic cement chemically melts the plastic it contacts, **9**. If it comes in contact with two or more bits of plastic touching each other, those surfaces will melt and fuse. They will not bond if there's a layer of paint between them. Always scrape any paint off surfaces to be bonded together.

I like liquid cement for situations where there may be a need for adjusting parts by sliding a bit, **10**, **11**, **12**. Testors liquid is slow to set and works well in this case. Other brands like Tenax-7R or Weldbond set much faster. I only occasionally apply the liquid glue with the applicator brush that comes in the bottle, preferring a small paintbrush. I had the misfortune once of dabbing the bottle's brush full of liquid onto a seam on a model, and didn't realize how much glue passed into the interior and pooled there. Only after the paint was

9 In the world of liquid plastic cement, Pro Weld (or Tenax-7R) cures quickly while the old stand-by, Testors, gives you more time to adjust joined parts and fix mistakes.

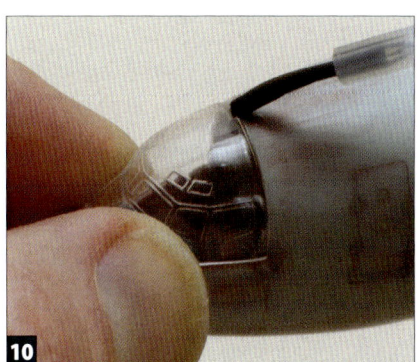

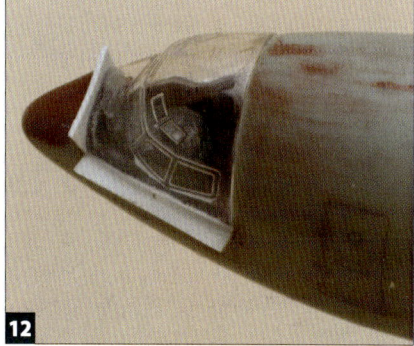

10 Here's a good example of where liquid cement is preferred. This canopy piece doesn't come close to clicking into place. The best match is to have the canopy follow the curved fuselage contours at the rear of the piece. A brush of liquid cement holds the piece in place and allows careful adjusting against the curved fuselage.

11 The piece is attached and fits reasonably well to part of the fuselage, but significant gaps remain on the bottom and front of the clear part.

12 These straight, narrow openings can easily be filled with sheet styrene. Had the piece been fitted to eliminate these gaps, the resulting mismatch at the curved rear of the piece at the fuselage would've been a big deal to correct.

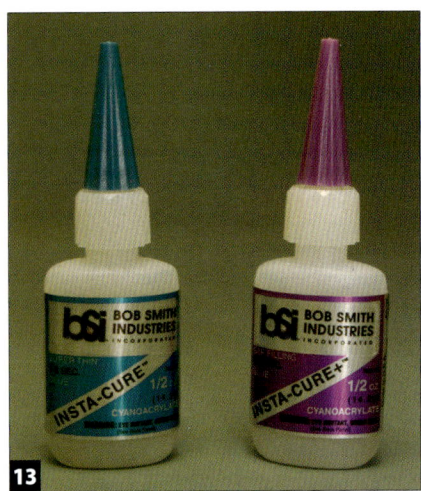

13 Two types of super glue—each with its own special applications—are the very-thin, fast-running type and the thicker, slow-running (and gap-filling) type.

on and I thought I was about finished did a corner of the airplane begin to wrinkle and then collapse, eaten up by glue from the inside. I still keep a couple different bottles of liquid cement on the bench, but overwhelmingly my first choice of adhesive for building models has become super glue (also called cyanoacrylate adhesive, or CA).

It really is super glue. Unlike plastic cement that welds parts together, super glue cures and holds pieces together without melting anything. Depending on the amount of super glue used, parts joined with it will be bonded forever. However you don't need much super glue to hold parts together securely. Herein lies one advantage of super glue. If you've ever needed to pry apart a model put together with plastic cement, you'll know that glued joints can sometimes be stronger than the surrounding styrene. If that happens, the only way you'll get some pieces apart is by literally tearing them, thus probably ruining them.

Super glue joints, on the other hand, can almost always be broken if needed with no damage to the part. The cured super glue can be scraped off the once-mated surfaces, and you're ready to try again. (Here too, you want plastic touching plastic without any paint barriers. A super glue bond to paint will be much stronger than a paint bond to the plastic, and the paint is likely to let go.) Another huge plus with super glue is the speed with which it cures. The recommendation found on kit instructions from the old days to "Allow assembled parts to dry (cure) overnight" today goes right out the window.

Super glue comes in a number of types. I use two for the most part. These are the standard liquid fast-running and the thicker slower-running "gap-filling" variety, **13**. I always buy small (¼-ounce) bottles and I've never used one up. Once open and kept at room temperature, super glue slowly thickens. As my thin fast-running type ages and thickens, it gets replaced with a new bottle. The old bottle becomes gap filling. As the proper gap-filling glue thickens to the consistency of molasses, it can really fill some gaps.

Because the thin runs so fast and you need so little, applying it can be tricky. I buy a length of 5mm-diameter nylon tube (found in hobby shops that cater to the RC crowd). I heat and stretch it as you would a piece of sprue to make antenna wire, **14**, **15**. Trim it down, jam the big end on the tip of the spout, and you've got a terrific fine point applicator, **16**. In time, it will clog on the inside near the tip. Simply trim off the end. When there's nothing left to trim, stretch another tube tip and you're back in business.

And fills gaps too. I rarely apply the gap-filling super glue straight from the bottle. I squirt out a blob, dip into it with a piece of stretched sprue, and apply it. A sharpened toothpick works just fine also. What I learned early on was that putting that blob of super glue on the bench at the same level as the parts you're working on is to risk minor disaster. I now fill a 35mm-film canister (pennies or BBs work) so it won't easily tip over and apply the blob to the plastic cap. What doesn't get used will cure on the flexible top and is easily scraped off.

I don't use much body putty as filler. When I do, it's the Tamiya brand. I fill seams, small gaps, and depressions with gap-filling super glue. It adheres to plastic in a permanent way putty never can. It does not shrink, does not crumble if you scribe a line over it, penetrates small cracks in a way putty can not, and bonds as well as fills. If I do use body putty, it's mostly to fill the scratches resulting from the coarse-grit sandpaper or sanding sticks employed against the super glue. But I often fill these lighter scratches with super glue as well and just sand it with a finer grit paper. Some may read these words in disbelief and horror, having heard that using super glue as filler is a recipe for disaster. Nonsense—you'll see.

A terrifically handy substance to have on hand is A + B epoxy putty (www.aplusbputty.com), **17**, **18**. Two equal parts are mashed together, resulting in

14 To make your own glue applicator, stretch sprue (or nylon tube as seen here). Hold the piece an inch or so over a candle flame and rotate it until it gets shiny and starts to sag.

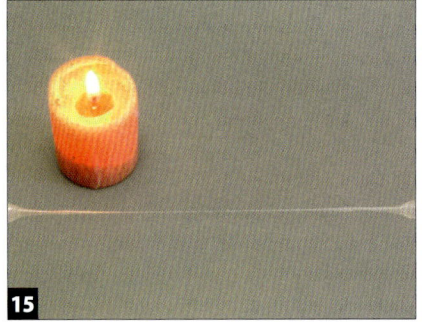

15 Remove the piece from the heat and pull the ends away from each other. With a little practice you'll get the feel of how much heat and how fast to pull to get any diameter you want.

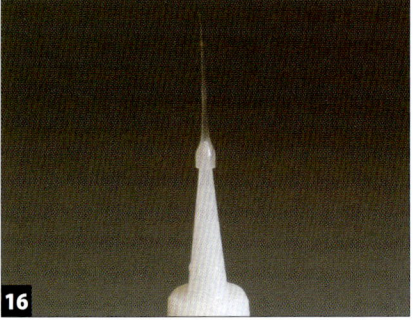

16 You can buy nylon spouts for applying super glue, but you can make a much more precise tip yourself. The trimmed heat-stretched nylon tube is pressed on the bottle's nozzle and is ready to go.

an extremely pliable, fine-grained putty that can be used to fill odd-shaped openings, beef up thin walls from the inside, or to mold or sculpt. It can be smoothed out wonderfully with a bit of saliva and a fingertip or the end of a cotton swab you've been sucking on. It cures rock hard (so be careful how much excess is left) and can be filed, sanded, and drilled. There's a "fast" variety that cures in about five minutes, but I don't find that product as smooth and easy to work. The regular putty cures faster with heat applied if you're in a big hurry, otherwise I give it 8 to 10 hours at room temperature.

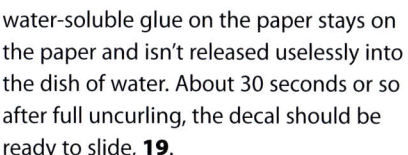

An extremely useful gap/hole/crevice-filling material is A + B epoxy putty. Mash together two equal amounts of the A and B and you create a pliable, fine-grained putty that cures rock hard.

It's an ideal filler in a case such as this, where you want to plug a deep, irregular opening.

Decals. Probably no aspect of finishing a model has more potential to provide joy or heartache than decaling. Good decals are a real treat to work with. Decals that don't conform to curves, break up as you attempt to slide them (or just refuse to slide), or are too big or too small are more than exasperating. While having things go wrong during your model project is never fun, at the decal stage, when the finish line is in sight, trouble is especially unwelcome. For that reason (and some hard experiences over time), I strongly suggest checking out your decal sheet (try one or two expendable items) before you get too far into your project. If you know ahead of time that you'll have to make substitutions, the blow is easier to take.

When cutting decals from the sheet, remove as much of the clear surrounding decal film as possible. Large simple shapes can be easily trimmed to their edges with good scissors. Smaller, more-intricate designs may require careful trimming with a No. 11 blade while the marking is still on the sheet. Precise knife cuts through the backing paper aren't necessary. Once the clear film is simply scored around the outlines of the markings, it can be plucked away with tweezers after the decal is wet and ready to slide off the paper.

To bring a decal to life, immerse it in warm water to which has been added a drop or two of dish soap. This acts as a wetting agent as it did in your "sludge wash" ingredients. As soon as the decal hits the water, the backing paper will start to curl. As the paper saturates with water, it will relax and uncurl. Generally, once the uncurling begins, I remove the decal from the water. This ensures that any of the water-soluble glue on the paper stays on the paper and isn't released uselessly into the dish of water. About 30 seconds or so after full uncurling, the decal should be ready to slide, **19**.

At this point, a dab of setting solution, if desired, should be applied to the location of the incoming decal. Different brands of decals vary considerably in their ability to snuggle down to the surface they're applied to or to conform to curves. Some absolutely refuse to do anything other than remain flat and stiff. This is where decal setting solution comes in. There are a number of these products on the market. They soften the decal itself and allow it to conform and cling to less-than-flat surfaces and help it settle into recessed areas as well. While they are all essentially vinegar/alcohol mixes, they vary in strengths, and reactions between different brands of decals and setting solutions are hard to predict. In some cases, I've seen no discernible reaction and in others, the decal is eaten alive. Good-bye decal. That's why it's a good idea to test the given setting solution and a sample decal from the sheet you intend to use before going too far. The setting solution may cause the decal to wrinkle considerably. Don't panic, but don't try to move it anymore either. You'll only ruin it at this stage. It will relax to its original shape by the time it dries.

With large decals, sliding the image off the backing paper directly on to the model is the fastest way to apply it. Smaller decals are more easily positioned by lifting them off the paper with tweezers and carrying them to their location on the model. Once in place, you may want to blot or wick away any excess water/setting solution with a gentle touch of paper towel or cotton swab. Removing this excess liquid reduces the decal's ability to move any farther. When the decals are totally dry, wiping away spots of dried glue and/or setting solution with a damp cloth will finish the operation.

Probably the worst thing you can discover is that due to age or bad storage conditions, your decals just break up when the water starts to bring the glue to life. There are a couple partial solutions. One is to brush a thin layer of Microscale's Liquid Decal Film over the decals you need to use. Another is to airbrush a layer or two of your favorite gloss coat over the sheet. In both cases, you're putting a fresh clear film over the fractured decals, which will bind the pieces together. These are partial solutions because while you can now at least get them off the paper in one piece, they probably won't be as flexible as you'd like. If they need to conform to a curved or complex surface, brace yourself for more frustration. Unless you have absolutely no other option than a defective sheet, I would pitch it and seek a plan B.

Another problem you may run into are decals that don't want to stick to the model. A layer of water-soluble glue is between the decal paper and the decal film. Sometimes the amount or the quality of that glue isn't sufficient. White glue diluted a bit in water can supplement what your sheet lacks. Some decal sheets (Tamiya and old Hasegawa come to mind) come with way more than enough glue. I save the scraps of especially "gluey" sheets to supplement less-endowed decal sheets.

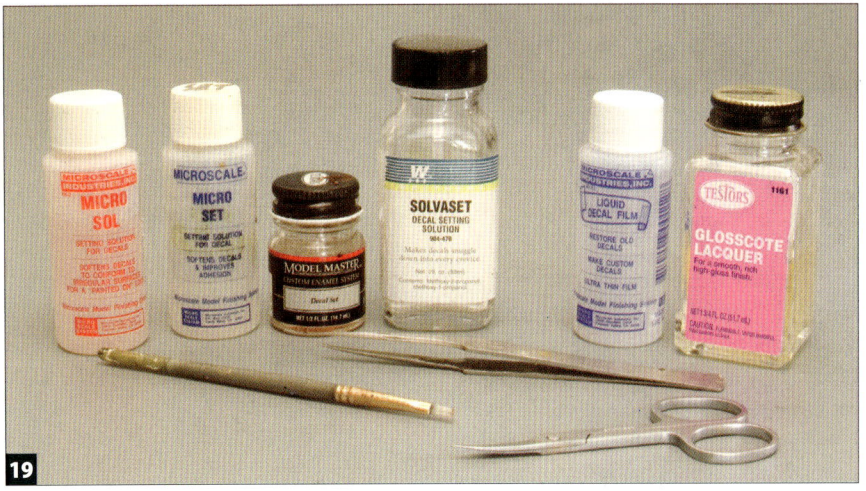

19 Here are some basic decal-application tools and products. Having a variety of setting solutions enables you to use one that is most compatible with the decals on your sheet. On the right are a couple fixes for old brittle sheets. You apply the film with a paint brush, the Glosscote with an airbrush. Tools are pretty simple: A short flat brush works well to smooth decals flat without tearing them. (I have one for decals only that I have shortened to set it apart from the rest of my brush collection). Pointy tweezers and good manicure scissors are valuable too.

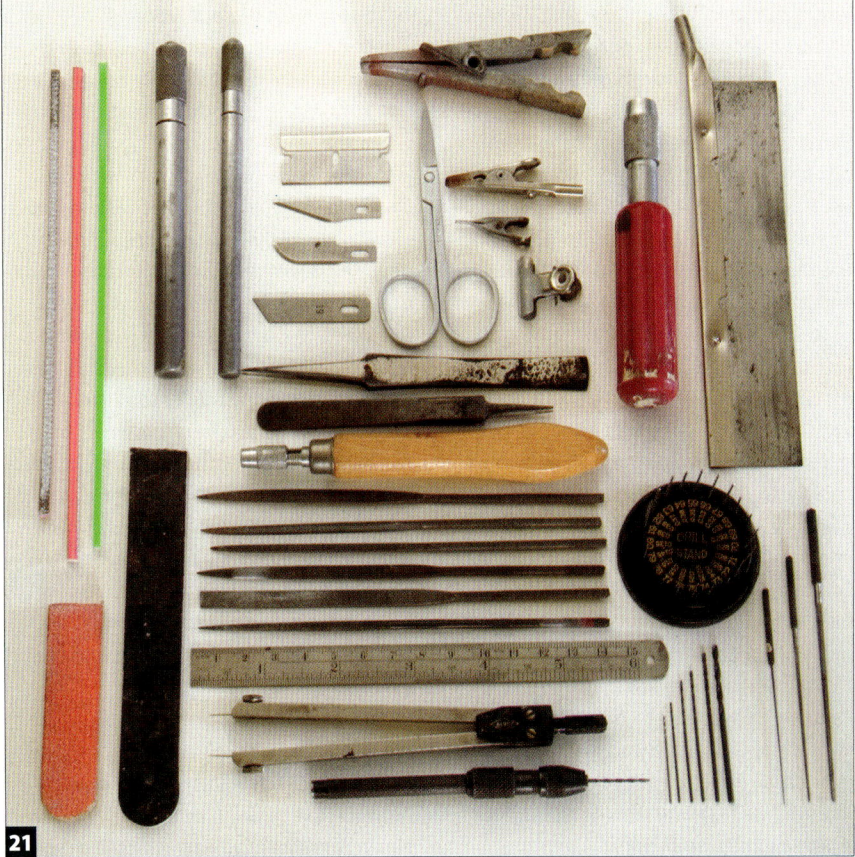

21 Essential tools on the workbench are sanding sticks, different-sized knife handles, an assortment of blades, a good pair of manicure scissors, a variety of clamps, files, a ruler/straightedge, dividers, drill and bits, saw with handle, and reamers in several sizes. Not shown, a punch-and-die set is indispensable for making tiny holes and discs, and with facing tape on sheet stock, perfect masking for rounded corners of canopy frames. Chrome tape punched on stock, makes round mirrors or landing lights. Also consider a scribing template and a drafting compass with a blade, not a lead, installed.

20 Here's why you don't want to put decals on a surface painted with a flat finish. Left is gloss, right is flat.

The most important aspect of making sure decals work well is to apply them on a gloss-painted (or coated) surface, **20**. The magnified surface of a flat finish resembles sandpaper. A decal applied to that will only settle on the high points of that sandpaper finish. When the decal dries, all the low spots of the paint job not touched by the decal will reflect light through any clear film. This will give you that most-unwanted frosted look. The magnified view of a smooth gloss finish is nearly level, so it connects with all of the undersurface of the decal, leaving no frosty voids under it so the clear film largely disappears.

Tools: Keep it simple. If you are brand new to this activity, plastic model building is about cutting parts, shaping parts, and connecting parts—maybe not in that order, but that's it in a nutshell.

Although you have your choice among a galaxy of cool specialized modeling tools, most everything needed to do first-class work is pretty unglamorous. A visit to any hobby shop or hardware store will provide the blades, saws, files, and abrasives like sandpaper you'll need to get started. I'll demonstrate a few exotic tools as we proceed with the projects in this book, but you'll see that for the most part, nothing fancy is needed.

Of course the longer you pursue the hobby, the more often you'll find the need for a tool you don't have. I suggest acquiring them as you go or as you find good deals. (You can always remind your significant other how much you would appreciate gift cards to miniature tool supply stores.) Scanning my own tool collection, which has grown over decades, I see quite a few obtained for a given situation that have rarely been touched since.

1/48 scale
SBD DAUNTLESS

Almost "out of the box"
4

This Monogram Dauntless demonstrates that building a kit almost entirely from stock components can result in a good-looking model.

The first project in this book reflects plastic modeling's original intent, which was to be fun. Monogram's 1/48th scale SBD Dauntless—a plastic dive bomber that could actually drop a bomb—was certainly a model I had fun with as a kid! Although it's a very basic model, for its age it's remarkably accurate in outline. Typical of old Monogram kits, everything fits pretty well, and with a little care, it can be turned into a respectable model. Let's build this kit close to "out-of-the-box" (or OOB or OTB), meaning no corrections or additions to what is supplied in the kit.

Remember, the point of this exercise is not to produce a great 1/48th Dauntless. There are much better kits to use as a starting point for that. Monogram's rendition serves here only to demonstrate simple model-building techniques.

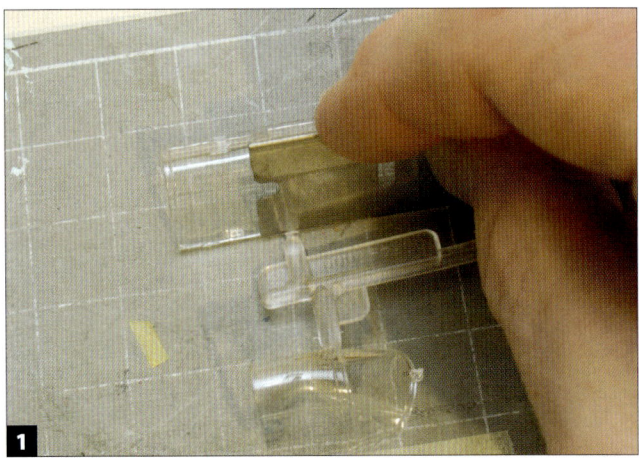

1 With a cushioned cutting mat underneath, a single-edge razor blade is applied to the sprue joint like a guillotine, making a perfectly clean separation of the canopy.

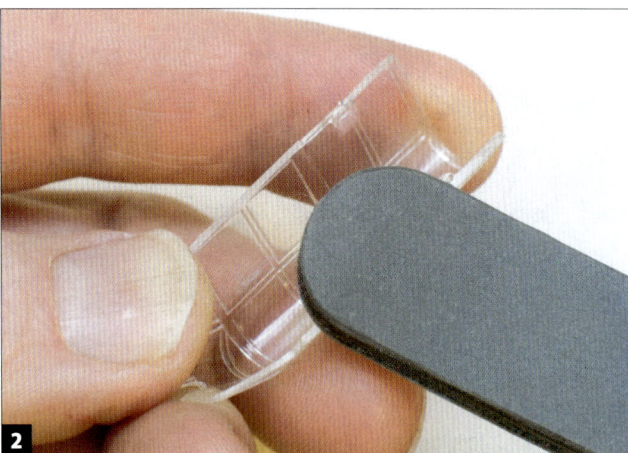

2 A fine-grit sanding stick smooths and levels the bottom edges of the canopy.

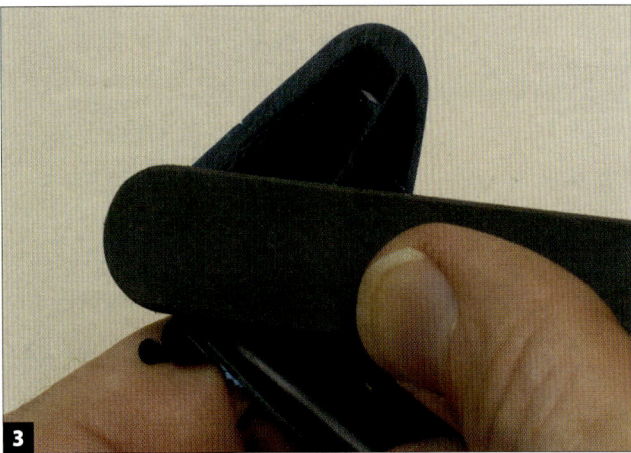

3 The same procedure is applied to all the mating surfaces of the kit, ensuring the tightest possible fit for all parts.

4 Test-fitting the canopy early in the building process alerts you to fit problems that you may encounter at the end of the process when the actual attachment of the clear parts takes place.

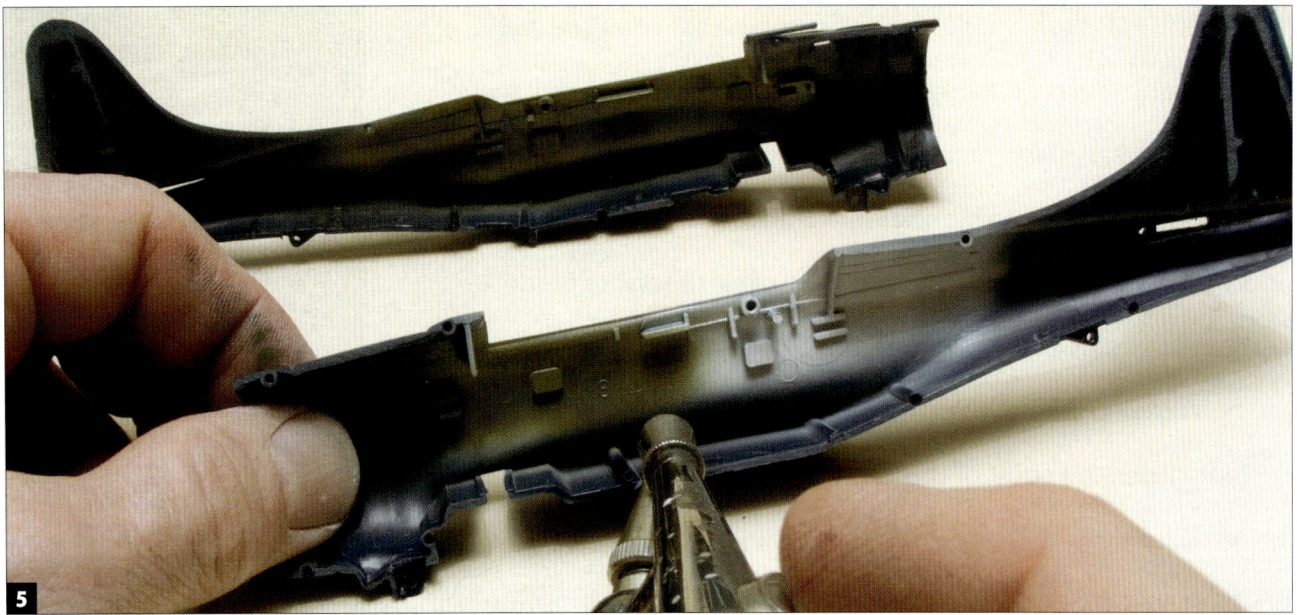

5 Following a base coat of white, interior green is airbrushed over the fuselage interior. Paint on the finger is evidence the photo isn't entirely staged.

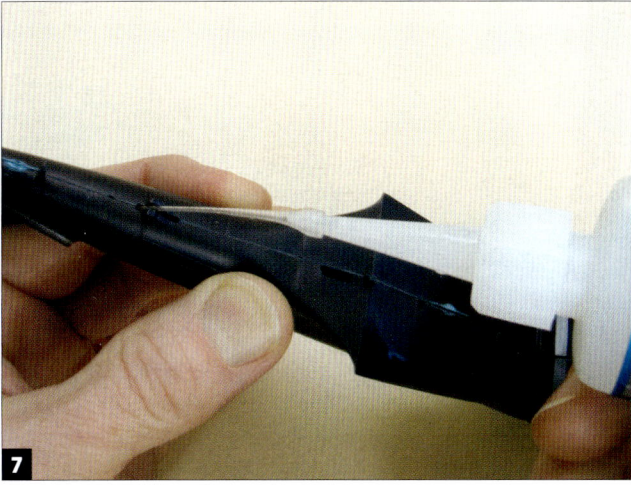

Using the stretched nylon nozzle, super glue is applied to the interior surfaces using any opening available.

With holes drilled to mark the front and rear limits of the cut to be made, a saw blade essentially connects the dots.

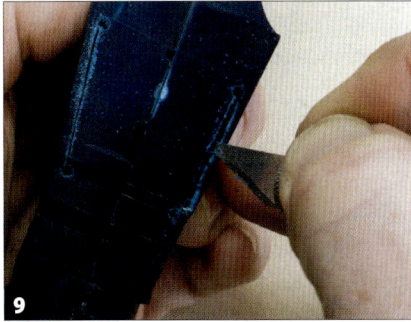

The resulting slot is carved open with a hobby knife with a No. 11 blade.

The upper wing half is held in place while liquid plastic cement is applied. The slow setting time of the cement allows any necessary adjustments.

Initial inspection. The first step is a parts examination. This doesn't take long on such a simple kit. I look to see that everything is present and free of noticeable warps, especially in the wing parts. Everything looks good and straight. Take a hard look at the canopy. It's in two pieces and this one is quite clear. Both pieces have tabs molded into the inside surfaces to ensure no lateral movement when fitting to the fuselage. Though clear, these tabs would be visible when the model is finished. Since this one is just for fun, I'm not going to worry about these tabs. Happily though, this kind of assembly "aid" is rarely seen anymore. The bottom edges also sport prominent ejector-pin marks (from the molding process) that need removal.

With a single-edge razor blade, cleanly separate the canopies from the sprues, **1**, then take a medium-grit sanding stick to all the edges, especially the raised pin marks, **2**, **3**. The trick in this operation is to remove only the raised imperfections and leave the edges of the part as intact as possible. Also, of course, try to avoid landing the sanding stick on the canopy itself. If you don't trust yourself for that, apply masking tape to the outside surfaces to protect them from slips of the abrasive. How the canopy fits its opening in the fuselage can cause unhappiness further up the road, so I want to know early on how well it will click into place, **4**. In order to know that, we need to match up the fuselage halves.

Fuselage has a fit. Snap the fuselage halves off the sprues and click them together to get a rough idea of fit; this one is quite good, revealing no serious gaps. Separate them, and with a medium-to-coarse sanding stick lightly rub the edges of the fuselage halves where they'll touch each other. This removes any mold imperfections (or mold release agents) that may keep the halves from fitting as tightly as they were meant to. It also provides a bit of "tooth" for the super glue. In many cases, I'll remove the small locating pins that fit into their corresponding holes on the other side of the fuselage. Building the way I do, you'll see they are not necessary. (Occasionally, you may find some of these molded-in helpers are actually a hindrance.) Edges thus prepped, dry-fit the fuselage and the clear parts. Happily, the clear parts fit better than expected.

In the spirit of keeping this project simple and fun, pre-assembly painting has been kept to a minimum, **5**. There is no cockpit detail to speak of, and that's OK. (Plenty of cockpit detail will come later in this book.) A closed canopy on this model won't invite a lot of inspection of the cockpit anyway. Study of some color photos of wartime SBDs suggests the interior color was yellowish olive green. For purposes of this project, I suggest you paint the cockpit walls and all the rest of the cockpit your own best guess at a mix of yellow-olive, and leave it at that.

Step one of the kit instructions calls for gluing a couple of interior pieces together. I have no problem with that. Step two is to

11 A strip of masking tape holds the dihedral of the wings in place. With the fuselage held in a vise at a true vertical attitude, the distance between ground and wing tips is measured. If the distances aren't equal, something is wrong somewhere, and an adjustment of the masking tape is probably in order.

12 With the proper dihedral ensured, the gaps at the wing roots get a squirt of super glue. This provides both strength and a degree of filler.

glue the antenna mast into the left half of the forward fuselage. Umm, I don't think so. Any bets on whether or not it will still be there by the time you're ready to paint? Things like this that can easily be knocked off should go on last. Besides, having to work around such a piece throughout a project is just annoying.

Modify the mount on the end of the mast so it passes through its slot in the fuselage. Eventually it will rest on the pin provided. The arrestor hook and bomb trapeze are likewise meant to be installed as the fuselage halves come together. For the same reasons as for the antenna mast, leave these items off (stored in a safe, visible place) until the model is mostly built.

Glue—one step at a time. Gluing the fuselage together starts with the tail. Make sure the two halves are perfectly lined up (and I mean perfectly), and allow a slight gap between the halves to open where you apply a small squirt of thin super glue, **6**. Press the gap closed. The glue flows by capillary action between the pieces and sets almost instantly.

Moving forward, handle each section the same way. Press the halves precisely together and apply the super glue from the inside, preferably. This keeps any potential mess on the inside where you don't have to do any cleanup. If you can't easily get to the inside, any opening where you can insert the glue spout is a good second choice, **7**. This includes rough spots like twisted-off sprue connections. Be careful not to let too much super glue loose at once. The stuff will pool in all sorts of places where you don't want it.

Pooled super glue does not want to stay in one place. If it should find your finger on a seam, you'll be lucky not to glue your finger to the fuselage, and at the very least, you'll have a super glue/fingerprint mess on the model to clean up.

Raised rivets raise their ugly heads. You'll always have a seam the length of the fuselage to deal with. Given the respectable fit of this kit and by ensuring the halves will be as tight as they can be before bringing them together, filling that seam should require minimal work. A complication here is all the raised rivets. It's nearly impossible to replace these tiny bumps if you sand them away, and a model festooned with this raised detail will look silly having a smooth spine and belly. The only solution is to protect the surfaces having raised details to whatever degree is practical. We'll deal with the fuselage seam after the wings are attached.

The traditional method of building a model airplane is to join the fuselage and wings together as completed sub-assemblies. This usually works fine but frequently results in a significant seam at the wing root. Filling and sanding will make the seam disappear, but in the case of this SBD, you've got these rivets to consider again. In the interest in minimizing a wing root gap and maintaining raised detail, we'll take a different approach.

Drill two holes in the bottom of the fuselage near where the wing will meet

13 An excess amount of wheel axle is given with the intent that this be melted flat, trapping a spinning wheel behind it (right). I trim mine flush with the wheel and fill any gaps with super glue (left).

14
The bottom wing half is attached and clamped in place with clothes pins.

15
A fine-grit Flix-I-File ribbon sander smooths the leading edge seam. Any further depressions found on the leading edge will be filled with gap-filling super glue and smoothed some more. Rivet loss is minimal.

16 A coarse-grit sanding stick grinds down the styrene plug that fills a slot in the fuselage. Sand until it is flush with the surrounding surface.

17 The marred surface reveals a low spot next to the plug.

18 A dot of gap-filling super glue on the end of a piece of stretched sprue fills the depression.

19 Some sanding of the area under the center dive brake reveals the left side to be seriously low. Since this will affect the trailing-edge fit of the dive brake, it needs to be dealt with.

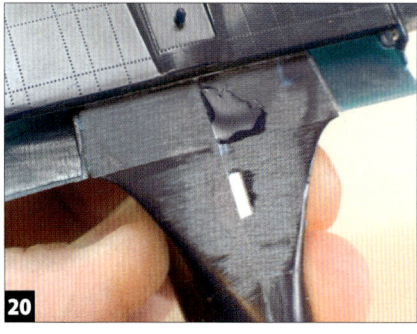

20 A puddle of super glue is applied to the depression. Below that, the glue-filled depression next to the styrene plug has been sanded smooth. Right of the plug, minor gaps are filled with thin super glue.

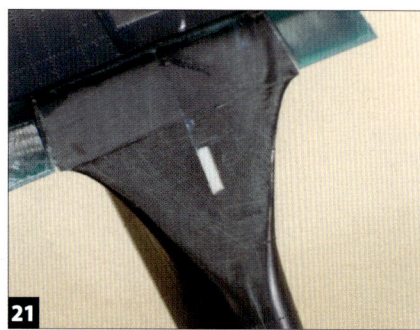

21 When the puddle cures (with the help of a little accelerator), more sanding-stick action is applied and the result is an even fuselage bottom area.

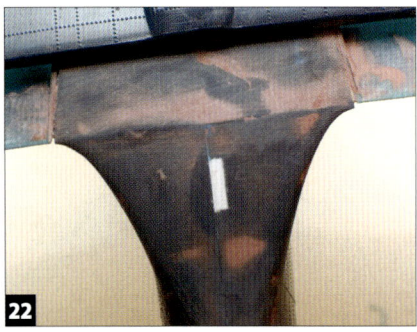

22 A heavy airbrushing of auto primer over the sanded areas fills scratches. The primer is wet-sanded away, leaving a surface ready for paint.

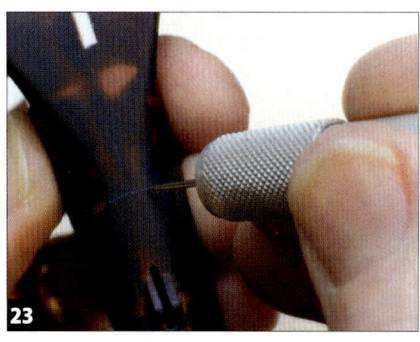

23 There's no way to totally avoid losing some rivets if you insist on no seams showing. To suggest some of the lost ones, simply poke holes with a needle. The raised ridge around the hole says "rivet" if you don't study them with a magnifying glass.

24 As for the rest of the fuselage seams, the tight fit of the parts means little fuss. A few swipes with a wet Flex-I-File ribbon eliminates most of the seam. Cracks that are still visible get a shot of super glue and the process is repeated.

it, **8**. With a saw blade in a hobby knife handle, saw a slot connecting the holes. Widen it enough to accept the ends of small clamps, **9**, **10**. With your sanding stick, rub the flat surface of the wing and the corresponding area of the fuselage, smoothing the surfaces that will touch. Clamp the wings to the fuselage and touch the seams with liquid cement. This allows you to adjust the wings to the fuselage to get the tightest possible fit.

Set the dihedral. To get the correct dihedral in the wings, stick a strip of masking tape to one wingtip, pull it across the top of the fuselage, and attach it to the other wingtip. Make sure there is enough tension across the tape to pull the wings up and against the fuselage. In order to make sure the wings are even, I put the tail of the model in a vise (making sure the vertical stabilizer is 90 degrees to the ground), **11**, and with a ruler, measure the distance from the ground to each wingtip. The distance must be the same on both. There will still be a bit of a gap at the wing root, but not a bad one. Fortunately, a black wing walk

25 The gun troughs on the upper cowling are tricky areas to smooth out as a seam runs through them. Sandpaper wrapped around a round file matches the contour of the depression and takes care of the seam.

26 Dry-fitting the dive brakes reveals some serious gaps between the brakes and the trailing edge of the wings.

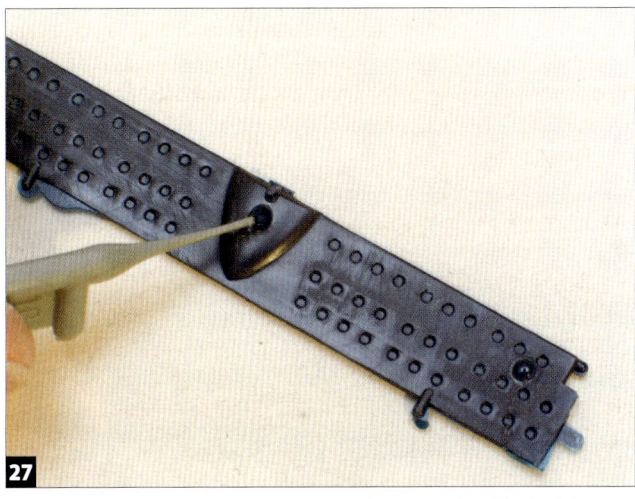

27 Any ejector-pin marks on exterior surfaces can be filled with gap-filling super glue.

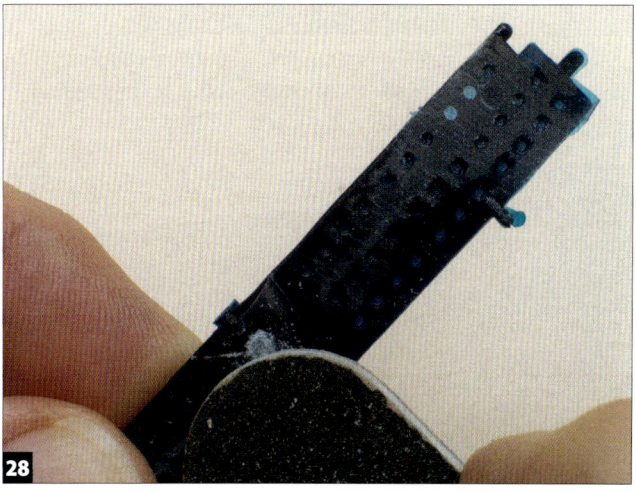

28 Preliminary smoothing of the cured super glue is done with a medium-grit sanding stick. This will be followed by wet-sanding with fine-grit sandpaper.

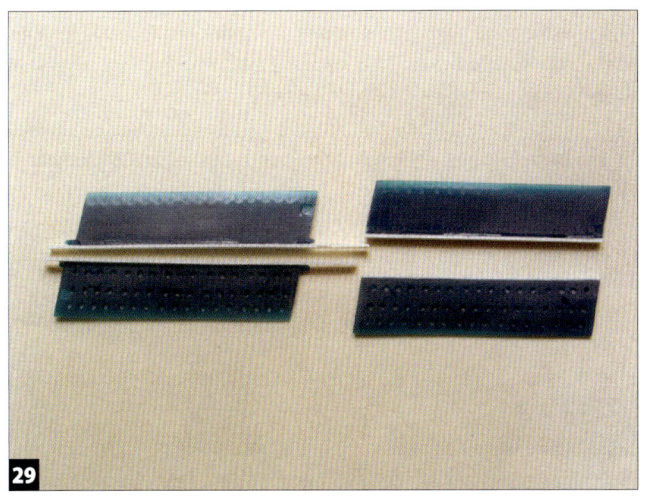

29 Strip styrene stock is glued to the leading edges of the dive brakes to close the gaps between them and the wing.

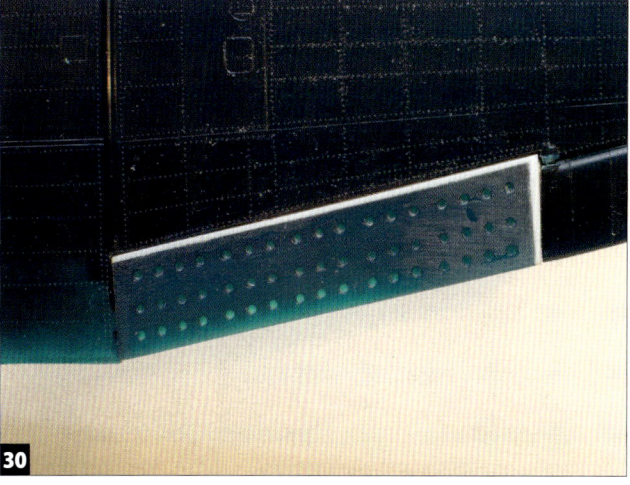

30 With two different styrene shims in place, the dive brake has a much tighter fit to the wing opening and fills it more realistically.

will use this gap as one of its borders, minimizing this seam's appearance even more. I carefully add a bit of super glue to this seam to fill it as much as I can, **12**.

Wheels up. The kit is engineered so that retractable gear struts are locked inside the wing halves. The landing gear is another one of those items that just naturally should be attached late in the process. However, with a nod toward nostalgia, I followed instructions and built my Dauntless with up-or-down landing gear. (This way I can at least retract them out of the way.) The axles receive a slight modification. According to the instructions, Monogram intended the wheels to spin (very important if you push your SBD across the floor, I suppose). The idea was to trap the wheels by melting the excess length of the axle against the wheel, **13**. Instead, I trim and smooth the excess length flush with the wheel. Then I fill around the axle with super glue and sand the wheel faces smooth.

With the upper wings fixed, it's time to test fit the bottom wing piece. If everything

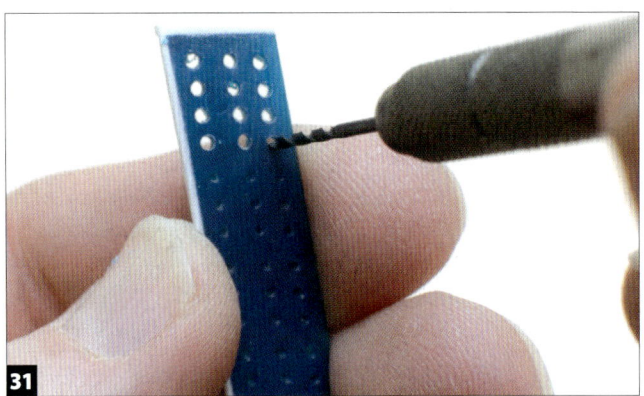

31 The tedious job of drilling out the dive brake holes begins. Rest assured that the work will be well worth it in the finished appearance of the model.

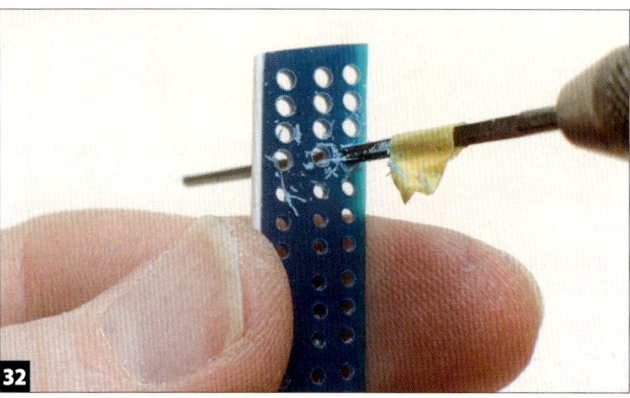

32 A reamer in a pin vise is marked with a piece of tape that says "not beyond this point." This ensures that the holes are all reamed to the exact same diameter.

33 Since the holes in the dive brakes will enable the insides of the wings to be seen to a degree, dab some dark gray paint on these areas to suggest depth, if they are visible through the holes.

34 The dive brakes are all shimmed, perforated, and glued in place. The styrene tube sleeve for the stand attachment is in the middle of the center dive brake. Also note the smoothed wheel covers—there's no sign of the axles through their centers.

35 The fit of the stabilizers to the fuselage is less than precise. Super glue fills the nasty gap.

36 A riffler file precisely grinds away excess super glue without destroying rivet detail. Masking tape protects the rivets.

37 The stabilizer joints are ready for priming. Note the wedge shim between the elevator and fuselage on the left.

lines up, the only seams you'll have to deal with are the wing leading edges. There aren't many rivets to worry about there, and the leading edge is easy to get to. As with the fuselage, start at one end, carefully aligning the wingtip halves and touching the end of the fast super glue spout briefly to the seam. With that end fixed, work your way to the other end, aligning, clamping, and gluing as you go, **14**. The super glue cures in less than a minute, and you'll be ready to work the leading edge seam with a fine Flex-I-File ribbon, **15**. Any gaps in the seam are filled with more super glue.

Seams again. With the wing and fuselage joined, consider the seams. The upper wing root gets a couple squirts of super glue to help fill the groove, but don't sand anything here. On the bottom of the model is a slot where a bomb trigger piece is meant to poke through. Jam a piece of scrap styrene into this hole, super glue it in place, and trim it nearly flush with the belly, **16**. Then come in with a sanding stick to wear it down truly flush.

In my case, the sanding stick revealed a low spot on the left side of the styrene plug, which I filled with gap-filling super glue, **17**, **18**. Further sanding revealed another low spot where the center dive brake goes, **19**. To ensure an even fit at the trailing edge of that dive brake, I filled this depression with super glue as well, **20**, **21**.

Some modelers have gotten the impression that super glue is a bad filler because it cures much harder than the surrounding styrene and will cause some kind of trouble when you try to sand it smooth. Yes, it does cure harder than the styrene, and it takes a coarse grit of sandpaper or sanding stick to wear it down. But until that hard lump of super glue is worn down some, the abrasive shouldn't be making much contact with the surrounding styrene. Use a coarse sanding stick in this operation but lighten your effort when you almost make contact with plastic. This is what I did sanding down the styrene plug.

Once the scratched-down super glue filler is almost flush with the surrounding styrene, wet-sand the area with finer grit paper (320 or 400) to smooth everything further. At this point, I recommend spraying the area with a filler/primer that fills

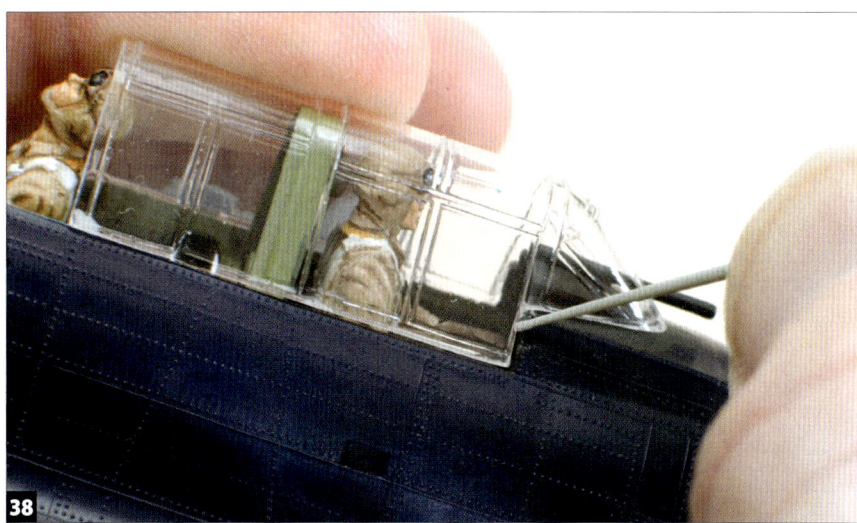

With the canopy pressed tightly down, small dabs of super glue are applied in a couple of the biggest gaps. This should hold the piece in place.

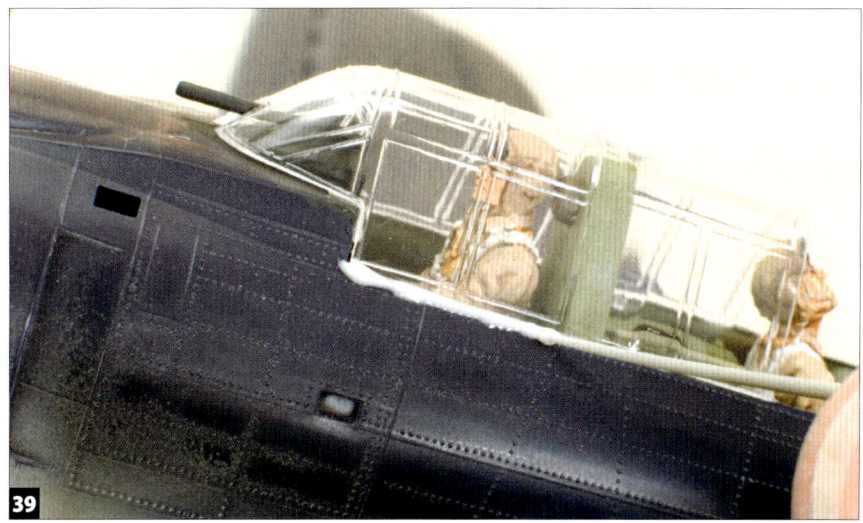

To preserve rivet detail, white glue is applied as a gap filler. Super glue and putty would ordinarily be used to blend at least the windscreen to the skin of the airplane. However, that means sanding, which would mean good-bye to raised detail.

Masking the canopy begins with pre-cut pieces of masking tape carefully laid in place. The rear gunner looks concerned. (Maybe it's good his gun hasn't been installed yet!)

41 All areas to remain clear are covered and protected. The open gunner's position is filled with a combination of tissue paper and Blue Tack. Slips of paper are slid between the dive brakes to keep paint overspray from getting through the holes and spoiling the red interiors.

42 An initial coat of interior green before the top color is applied ensures that the framing seen on the interior is green and not blue-gray.

scratches and reveals any pits or gaps that weren't obvious beforehand, **22**. Then wet-sand with 800- or 1000-grit sandpaper, and unless a flaw shows itself, the area should be ready for paint.

The rear fuselage seam disappears with a little wet sanding with a Flex-I-File, **23**. At this stage, some rivets had to be sacrificed. To restore those few that got sanded away, I suggested them with pokes of a needle chucked in a pin vise, **24**. The vise gives you great control over the needle, which leaves indentions in the styrene surrounded by raised ridges. Unless checked under a magnifying glass, these ridges look very much like raised rivets.

The separate cowl piece has a complicated top cross section to match the fuselage. Between the gun troughs and the raised air scoop, there are some slight mismatches. Once again, dab some super glue on the trouble spots and carefully sand the areas to create unbroken contours. To get the smooth curved recesses of the gun troughs, wrap some 400-grit sandpaper around the handle of a file and use it as a curved sanding block, **25**.

Brake job. The one significant modification I'll make to this kit is improving the dive brakes. Their "Swiss cheese" look was a distinct characteristic of the SBD, but Monogram's treatment of them was simplistic; instead of holes, you get dimples. They were also part of the play value of the kit—they're hinged to open and close.

As with most moving parts on a model, a good amount of slop had to be engineered into the fit of these dive brakes so they move freely. The first part of this job will be to eliminate that particular feature and tighten them up, **26**.

Fill any ejection pin marks with super glue and sand and smooth them, **27**, **28**. Next, glue strips of .020" sheet styrene on the brake/flap leading edges. Trim and sand the shims even with the brake/flap surfaces, filling any gaps with a bit of super glue, **29**. Test-fitting the pieces into their respective sections of the wings helps determine how much shimming is required on the ends. Add these extensions in the same manner as the first ones, **30**.

Drill, drill, drill. Drill pilot holes for the dimples, all 268 of them, with a $1/32$" drill bit in a pin vise, **31**. Where's the Dremel tool? In this case, I want the control offered by a hand tool. I've never melted a piece of plastic through friction created by a hand tool, but I can't say the same of a motor tool.

Once all the dimples have small pilot holes, we need to enlarge them. A $1/16$" drill bit will open them up them instantly but may chip edges. A reamer will open them without any nicks but will take at least twice as long. Let's go with the drill bit in the interest of speed and because this is just a demo model. A few chipped dive brake holes won't be distracting when the model is displayed. If you try the reamer, mark the point on the reamer that creates the proper-size hole with a bit of tape, **32**. Boring all the holes to this marked point ensures they will all be the same diameter.

Unlike nearly every other SBD model I've seen with this option, the dive brakes on ours will be closed. That's because this is a quick "build" just for fun. Besides, having closed dive brakes is a far more-typical configuration. First, brush-paint some dark gray on to the flat trailing edges of the wings and the fuselage center section, **33**. This is to suggest some depth and shadow. Just as with the wings and fuselage, start by fixing one end of each brake/flap where it belongs and work your way to the other end, adjusting and tacking in place the piece as you go. They are all thin and a bit warped, so all need a bit of attention.

At this stage, start thinking about a gear-up display option. The landing gear will be retractable, and the interior detail will consist of a couple of crew figures. This is a natural for an in-flight model. To accommodate a pin in your display stand, drill a hole through the center dive brake and underlying fuselage and insert a section of styrene tube. This will serve as a sleeve, **34**.

Meet the riffler file. The last additions that make the Dauntless shape complete are the horizontal stabilizers. As with the top wing pieces, smoothing the bonding areas of both the stabilizers and fuselage are in order. Even doing that, the match of the parts isn't great, with significant gaps. Bring on the thick super glue one more time, **35**.

When cured, use a riffler file to grind the glue and some of the surrounding plastic smooth, **36**. (Rifflers are curved files that allow you get into tight places that straight files cannot go.) On my model I needed to dab super glue into low areas four times until everything was acceptably level, **37**. Medium grit (320) wet sanding is accomplished with masking tape applied to protect the rivets. This isn't a perfect result, but it's good enough for this project. Note the lines of rivets running not quite parallel to the fuselage.

Crew change. The last step before painting is attaching the canopy, but we can't do that until the interior is complete. Glue the decal instrument panel in place, and then think about how to handle the crew. The kit comes with a pilot and a gunner figure, but the pilot in particular doesn't look too great. Although I like the goggles pulled down on his eyes, the head is poorly shaped. As much as I want to keep this model stock, I replaced my crew with more-modern and better-sculpted airmen from a Monogram PBY kit.

The canopy fits adequately, although it has some gaps. With small amounts of super glue on the end of stretched sprue, fill some of these gaps and fix the canopy in place, **38**. The coating of Future prevents the super glue vapors from frosting the clear part. The unsightly gap along the length of the canopy could also be filled with super glue, but because of the rivet detail near this gap I chose Elmer's white glue instead, **39**. Again, with a stretched bit of sprue, I spread a bead along the gap. When this dries, a wet cotton swab rubbed along the joint removes any glue not right in the gap. With the excess wiped away, the rivets remain and any gap is filled. Although it works in this instance, this method of canopy gap filling does not hold up to close scrutiny, but this model is about learning, not about close scrutiny.

Canopy frames—no escape. Masking canopy frames is one of those exercises hardly anybody likes. There's no quick, easy, secret way to get the perfectly sharp, straight lines required. This is one step in the process that simply requires focused

43 The national insignia decals are trimmed as close to the image as possible, eliminating the surrounding clear film.

44 Solvaset is applied to the decal location. The model's gloss paint eliminates the need for any gloss coating before the decal stage. Note the flat-black painted wing walk and red interior of the dive brakes showing through the many holes.

45 Once the decal is in place, add another application of Solvaset around the edges. This ensures that the decal will conform to surface details.

Here is the result of the Solvaset action on the insignia decals. Mostly the results are what was hoped for, though that desired "painted on" look wasn't quite achieved. Bubbles formed at the intersections of some of the rivet lines. Thinner decals or fewer rivets would make the difference. Note the tight fit and even holes in the dive brakes.

Decals that meet on a trailing edge, like these rudder stripes, rarely give a good accounting of themselves. Trim any excess decal film and touch up the un-decaled edge with paint on a fine brush. Note the sludge wash applied to the trim tab and elevator hinge lines.

attention and precision. Aftermarket "easy masks" of pre-cut vinyl are a recent development, and the theory behind them is a good one, but my limited experience with them has left me unimpressed. They have not given me the perfect fit that cutting my own masks provides.

In the case of the SBD, the canopy is one of those small "greenhouse" affairs with significant framing. The good news is that the majority of it is simple, straight, and square. Here's how to handle the frame: Start by laying a strip of Tamiya masking tape (very thin and flexible) on a cutting mat with grid lines marked. Cut this strip lengthwise a few times using the lines on the mat as a guide. Cut each strip into many short lengths, making careful 90-degree cuts, again using the lines on the mat as a guide. Perfectly square ends on the pieces of tape is the goal.

Once the pieces of tape are cut, pick up your best pointed-end tweezers. Peel up a strip of tape and place it on one of the window panels, making sure the square end matches the square corner of the tiny windowpane, **40**. If the strips of tape are truly square and you get the corners down where they belong, the long sides of the windows will automatically be covered completely. Repeat this procedure until the windows are covered. The windscreen with its curves is a little trickier. Here I just cut curved pieces of tape with the help of any curved template in my collection.

There's much more trial-and-error fitting with curved pieces.

To plug the cavity of the gunner's position, stuff a small piece of paper towel into the opening, **41**. To seal the edges—especially around the canopy—press in small amounts of Blue Tack with a toothpick. Blue Tack (other brands are also available) is a tacky putty marketed for hanging posters on walls. It doesn't affect paint and is handy for stuffing holes like wheel wells prior to painting. In the case of this canopy, it will provide a seal against anything forced in by airbrush air pressure.

Once the clear areas and cockpit interior are covered and protected, spray the exposed frames with a chromate green, **42**. This way the frames will be green on the inside and blue-gray on the outside.

Paint shop. The SBD paint scheme is simple—gloss light gull gray on the bottom and gloss blue-gray on the top. Apply a blue-gray base coat on the upper surfaces, then add a bit of black to the mix, and airbrush random patches of this new shade on the top surfaces. This gives an uneven look to an otherwise too-even finish.

Next, alter the mix to a slightly lighter shade by adding a little tan and lightly spray the fabric of the elevators and ailerons. I also add a little random mottling to the rest of the airplane with this lighter mix. (After many years of studying photos of real warplanes, especially those with a single-color upper surface, what strikes me is how over time the finish becomes darker here and lighter there. These patterns may be due to where and how the ground crew worked on the plane. Regardless, I just tried to duplicate an "uneven" lived-in look on the Dauntless.)

Decal time. After waiting a day for the paint to dry, it's time to go to the decal stage. The kit decals look thick but usable. Though not really accurate for a World War II SBD (the markings in this issue represent a restored airplane), they're close enough for this project. Trim all the images as close as possible to eliminate the surrounding clear film, **43**. Since they will be applied over all those wonderful rivets, prepare the area with Solvaset, **44**, which is a fairly strong decal setting solution. Normally Solvaset is my last-resort decal setting solution, as I've found it to be the most powerful. In this case, the thickness of the decals, combined with the raised detail they will by applied over, suggest reaching directly for the Solvaset. Once the decal is on the wing, add a bit more Solvaset around all the edges, **45**.

The decals will dry in four or five hours. Where they are stretched over things like the slats near the wing leading edge and the aileron hinge lines, score them with a fresh No. 11 blade and brush on a bit more Solvaset. This helps them snuggle down into these deep recesses. Do the same where there are tiny elongated bubbles over some groups of rivets. The result is not perfect but good enough for this project, **46**. Just one more reason not to like raised rivets.

The rudder stripes are also decals, and—as is typical—they don't wrap around the trailing edge of the rudder very well. After trimming away the excess film, dab the length of the trailing edge with white paint, **47**. When this dries, go back with red and fill in the missing ends of the stripes. This is the first of a lot of paint brush tweaking that this (and every) model will receive.

Finally, apply the same dark gray "sludge wash" as applied to interior areas, to all the distinct recessed areas such as the control surface hinge lines, step cutouts in the fuselage, and cowl flaps to suggest depth. When this dries, wipe down the model with a damp cloth to clean up excess sludge wash, decal glue, and Solvaset residue. This is followed by a close inspection to make absolutely sure all traces of anything other than paint and decals are gone. Airbrush the model with several thin coats of Testor's Flat Clear Lacquer. Lightly applied, it dries almost instantly. (And lightly is the only way you want to apply it on enamels.)

Moment of truth. With the gloss finish dulled down, the moment of truth arrives for the canopy. In this case, the inside stayed free of any unwanted spray, and the masking tape did its job, remaining pressed down tightly around the frames, **48**. Revealing those shiny clear windows surrounded by sharp, flat framing is always a moment of excitement for me.

We've nearly reached the finish line. In order to make the finished piece look like a miniature airplane and not just a nice model, I studied photos of early wartime SBDs. What do the photos show that my model does not have? Mostly I noticed smudged areas where oil, grease, or fuel may have been spilled or smeared around. With a pointed paintbrush, lightly apply dabs of dark gray chalk pastel mostly around the cowling and cockpit and the center section of the wing. This is where the vast majority of the human activity on the airplane would take place. What the photos didn't show was any significant chipped paint. This makes sense as

With the flat coat applied, the masking tape can be pulled off the canopy. Note the paper towel between fingers and the model at the rear of the fuselage. At these last stages, bare fingers on a model are never a good thing.

After attaching the sprue line to the top of the antenna mast, it is stretched to touch the dab of super glue on the tip of the mast on the stabilizer. It's pulled taut, but not too taut.

There's a fair amount of slack in the wire, but this is easily fixed. Note the visible film around the fuselage codes where the decal didn't settle over the rivets. A bit of close touch up work with a fine paintbrush will cover the frosted film.

51 The glowing ember of an extinguished toothpick is positioned under the sprue at this distance underneath it, and …

52 … bink! The sprue line snaps taut.

53 The secondary wire is attached to the primary wire with a tiny dab of white glue. Note the lack of any gaps between the canopy and the fuselage. You should never see a gap between canopy and fuselage. Real airplanes aren't built that way.

exposed aluminum on carrier-based aircraft is tended to quickly before salt attacks it.

Back to step 1. Now that most of the handling is done, we can return to that antenna mast and the pitot tube. It's a simple matter to glue them in place now.

There's also the antenna wire to tend to. Heat a length of black sprue over a candle flame. When it gets shiny and begins to sag, pull on both ends, resulting in a long filament. The diameter of this filament will vary, so inspect it for a suitable length that's thin and even. Add a post (also made from a piece of stretched sprue) on the tip of the tail.

Touch one end of this stretched antenna filament in a drop of super glue and touch it to the tip of the antenna mast. When it dries, apply a small dab of super glue to the post on the tail, **49**, **50**. Lightly pull the unattached end of the sprue to that super glue and allow it to touch.

OK, so now there's a wire, but it's anything but straight and taut as it would be on a real airplane. Light the end of a toothpick, allow it to burn a little bit, then blow it out. With the still-glowing end, approach the wire from underneath, **51**. You'll know the toothpick is close enough when the wire starts to jiggle—it will then snap taut, **52**. Be ready to yank the ember away before anything else bad happens (and anything else would be bad). A single-edge razor blade makes a nice guillotine cut on the end of the sprue extending from the post on the tail.

Like many aircraft, the SBD also had a secondary wire from the main wire leading into the fuselage. Touch one end of a short length of sprue filament the same diameter as the main wire to a dab of white glue, **53**. (Use white glue, because if the placement of this wire isn't perfect, the white glue will release cleanly and you can try again.) Touch the existing wire with a spot of glue and place the secondary wire there with a tweezers. This glue dries quickly, and the other end of the second wire can be attached where it belongs with white glue or super glue. Trim the excess with the razor blade.

There you have it—a straight-forward "build" of a handsome model plane. I hope you learned something in the process of modeling this old classic.

2-B-4

Douglas
SBD-3
"Dauntless"

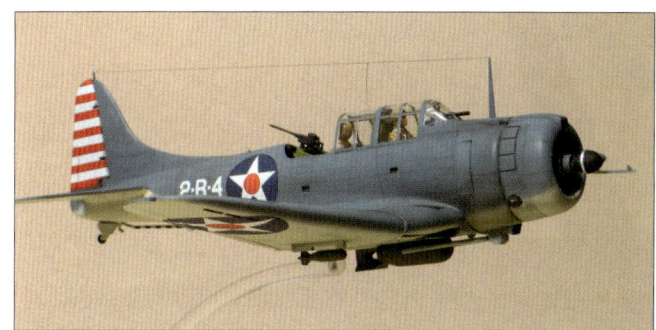

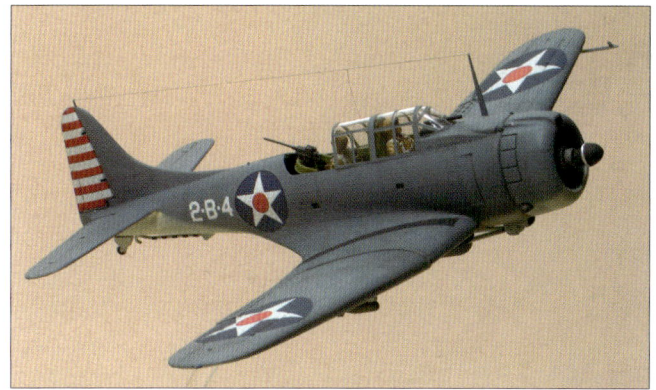

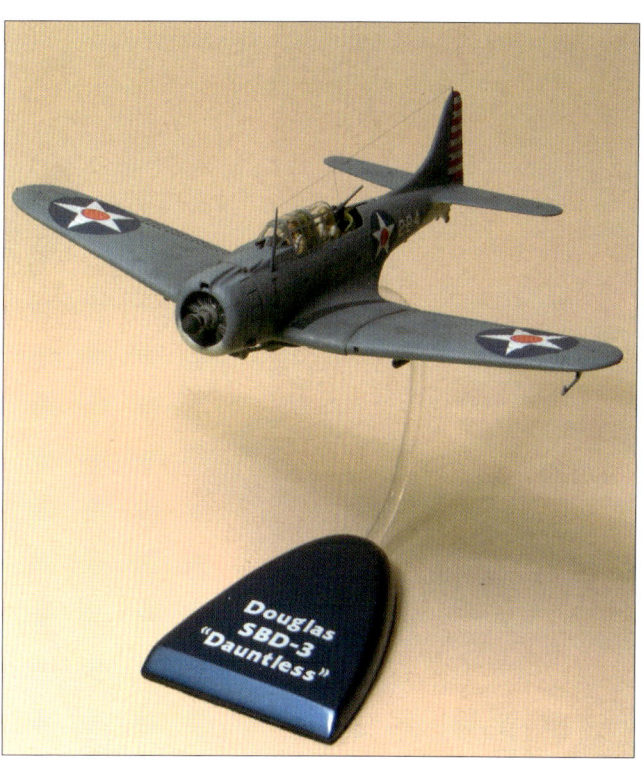

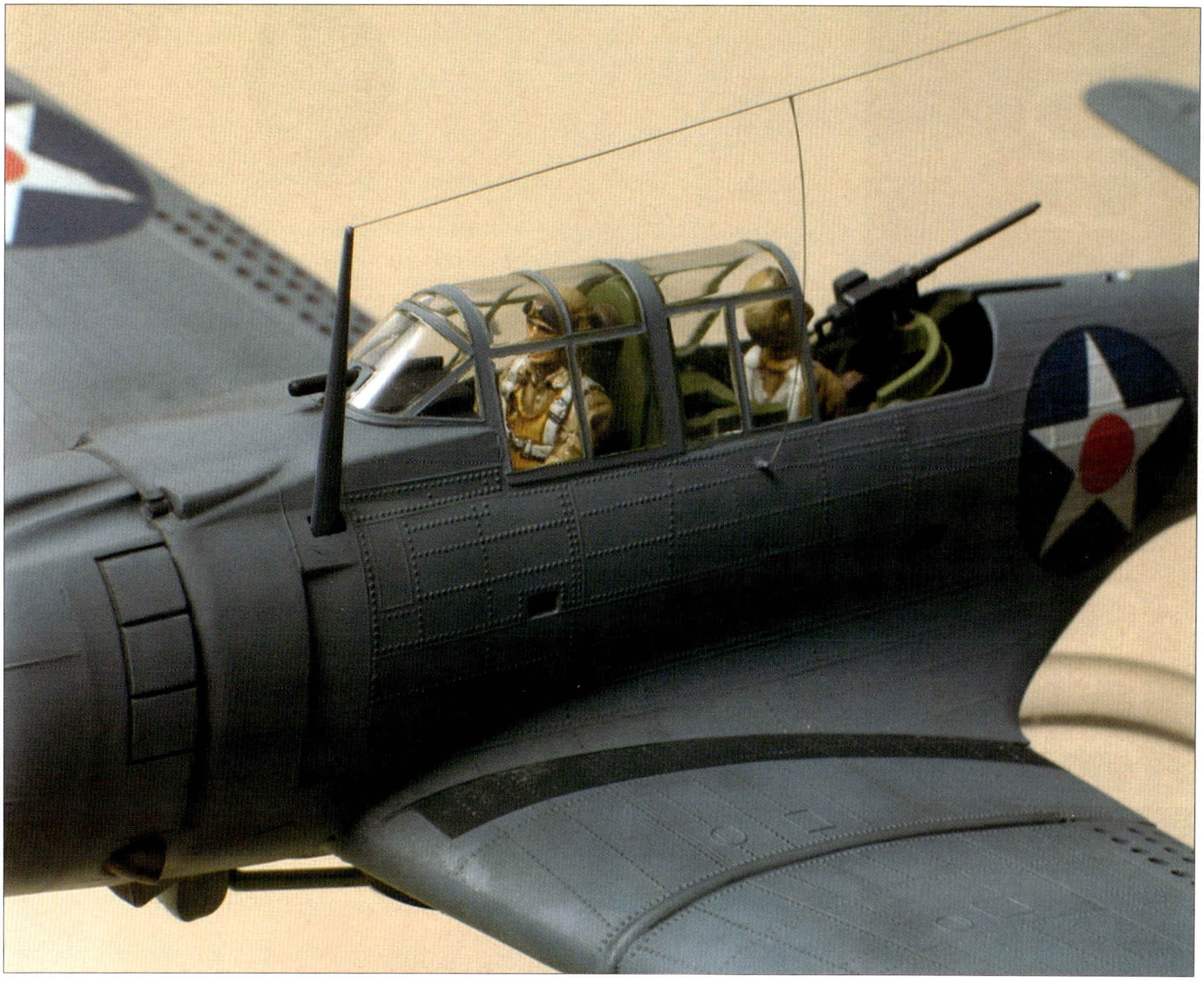

5

Enhance a modern kit

While the SBD of the previous chapter represents a state-of-the-art kit from the early 1960s, this project is closer to the standards of today. Released in 1995, Tamiya's P-51B is a terrific kit. In 1996, Tamiya released the same kit as an RAF Mustang III, which featured the blown "Malcolm Hood" (canopy) you see here.

The Tamiya Mustang falls into the category many modelers call "shake and bake kits," meaning the fit is so good, you can almost throw in some glue, shake the box, and the pieces will practically fall together by themselves.

The most obvious difference between the Monogram SBD and the Tamiya P-51B is the beautiful recessed panel line detail on the Mustang. It's a great kit, but even a great kit can be improved.

1/48 scale
P-51B MUSTANG

Homemade seatbelts, buckles, and harnesses help detail the cockpit of this Mustang. Modeling a closed-canopy plane like this one entails some different techniques than modeling a plane with an open canopy and crew figures.

1 A drafting eraser shield is used as a straightedge for cutting lead foil. Using a hobby knife handle like a rolling pin, lead foil can be pressed and rolled quite thin.

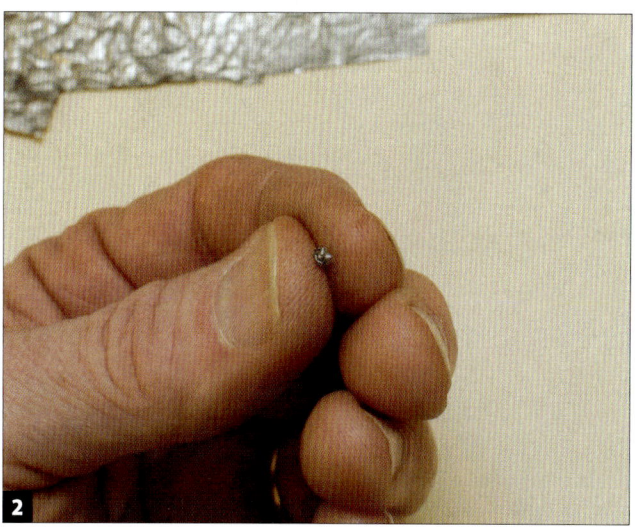

2 Once the strips are cut, they are rolled and compressed into tiny lead balls.

3 Carefully unroll the lead balls with tweezers to give the strips a wrinkled texture that better simulates 1/48 scale canvas straps.

4

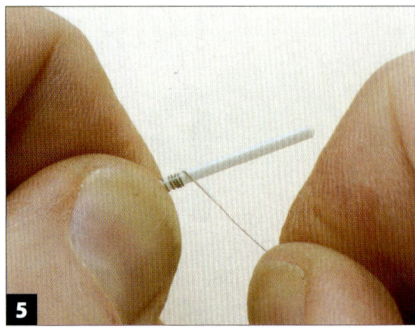

5 Tightly wrap fine wire around a piece of Evergreen square strip styrene.

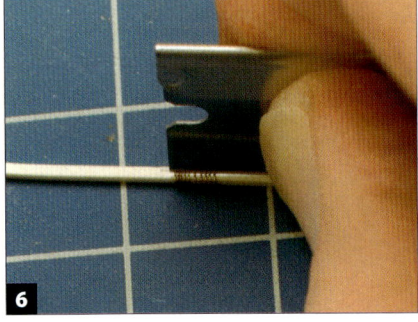

6 A razor blade works well for cutting the wire on one side of the strip. The result is a bunch of wire square "buckles."

7 The cut loosens the squares enough to be coaxed off the strip stock with fine tweezers. These same tweezers will be used to force the cut ends back to touching each other, reforming visually intact squares.

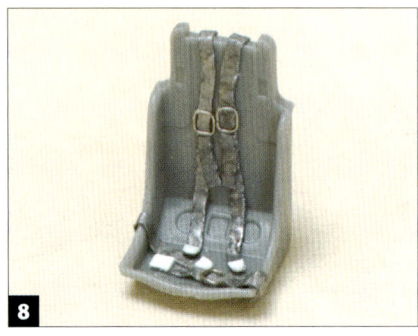

8 The seat, with its new belts and buckles, is now ready for a paint job.

9 The cockpit floor assembly is ready to install. The smooth plywood floor was generally painted black. Wear is simulated with dry-brush dabbing of flat tan.

Buckle up. For one thing, the seat needs harnesses. These are easily made from lead foil, laid flat and cut to width (in this case, 1/32") with the help of a metal straightedge, **1**. Ball up these pliable strips between your fingertips, then carefully unroll them to create believable folds, **2**, **3**, **4**. A length of thin wire—about like carpet thread—is tightly wrapped around a square of .030" styrene stock to make buckles, **5**. To get them off the plastic, make a cut with a single-edge razor blade while guiding the tiny wire squares with fine tweezers, **6**, **7**.

Superglue the wrinkled lead foil strips to the seat and add the wire tightening buckles on top of them, **8**. Small shaped bits of styrene strip added to the ends of the straps suggest other belt hardware. With the addition of these buckles and a little attention paid to painting, the seat can easily become the focal point of this interior, **9**.

Easy exhaust. Another small touch that adds interest is to drill out the exhaust stacks and gun barrels, **10**, **11**. In 1/48 scale, this isn't too demanding. All it

INSTALLING AN AFTERMARKET HARNESS

Photo-etched belt and harness buckles are available, and in scales bigger than 1/48th, you'll want to strongly consider using such aftermarket parts. Like threading a needle, you feed your lead foil strip through the tightening buckles, and wrap the end through the slot in the belt fasteners on the ends of the strips. For the purpose of illustration, here's a set of 1/32 scale belts attached to a Hasegawa Spitfire MkV seat.

Actual photo-etched belt buckles are available for those with good eyes and steady hands. Here a strip of lead foil is threaded through the thin slots in the brass pieces.

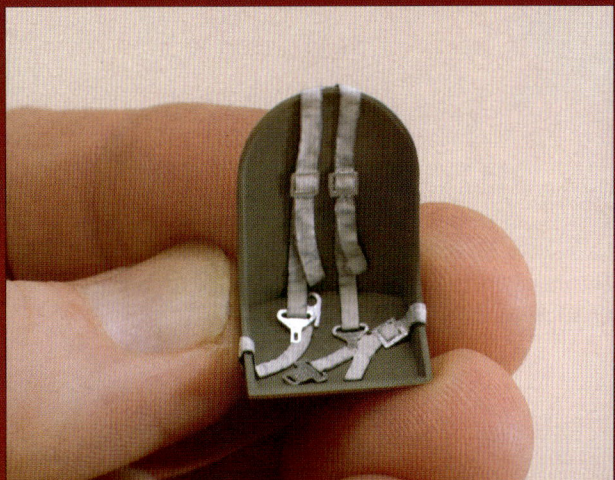

There's nothing wrong with this seat and its belts, but the appearance is rather flat.

To add some depth, first draw a line of black artist's oil paint next to a belt with a fine pointed brush.

Pull the paint away from the belt with a wider "blending" brush. The idea is to blend the outside edge of the oil paint.

You can do this anyplace you find a raised edge or a corner. The result is a realistic shadow effect.

10 To drill out the fronts of exhaust stacks or gun barrels, first press a pilot hole into the center with a needle or a scribing tool.

11 With that hole as close to dead center as possible, slowly drill it out with a small bit in a pin vise.

12 A piece of masking tape is rolled and placed sticky side up on a cutting pad. Press the stabilizer on the tape to prevent it from moving as a needle scores the line between stab and elevator.

13 That handy single-edge razor blade is used to cleanly slice through the thin outboard edge.

14, **15** Stubby tweezers grip the "ear" of the elevator after sufficient scribing. Wiggle and twist the piece until the remaining plastic fatigues and the ear comes loose.

requires is a steady hand and some small-diameter drill bits. If you don't want to drill, a dab of flat black paint on the ends works too.

(And before you paint those stacks "rust" as most kit instructions suggest, find a good color photo of these objects or examine a real piston-engine airplane.

An old airplane mechanic told me that exhaust stacks are usually stainless steel. You'll find heat discoloration, but probably not a color you'd describe as "rust.")

Falling elevators. In order to add a little more realism to the model, let's drop the elevators. Most photos of parked airplanes reveal the elevators (and often the rudders) to be slightly out of the neutral position in which plastic models are almost always molded.

To duplicate the relaxed position gravity imposes on the elevators, start by scoring the hinge line, **12**, **13**. Any sharp pointed tool will suffice. Make as

THREE PATHS TO BETTER INSTRUMENT PANELS

The other point of focus of this interior (and all interiors, really) is the instrument panel. There's something captivating about that collection of dials and gauges.

Tamiya molded this Mustang's instrument panel with raised dial detail, but also provided a decal if you don't want to tackle the instruments with your own paintbrush. However, the raised detail isn't that much of an improvement over the decal. For this model, I painted the panel myself, and for the sake of illustration, made up another one by shaving the raised detail and applying the supplied decal. And just for fun, I painted up a Monogram P-51D instrument panel for comparison. (In my opinion, Monogram does this sort of raised detail better than anybody.)

All are first painted flat dark gray and dry-brushed silver. On the Tamiya panel installed in the model, I filled the instrument faces with gloss black paint. On the one I decaled, I had to trim the decal down a bit for a good fit. With the Monogram piece, I also filled the instrument faces with gloss black, then picked out the raised detail in the dials with a Prismacolor pencil (No. 914 "Crème"; white was too stark) sharpened to a needle point.

On the far left is the Tamiya piece with the supplied decal applied after the raised dial detail has been shaved and scraped off. The center example is the Tamiya piece without the decal, with the raised dial detail intact and filled with gloss black paint. At right is the panel from the Revell-Monogram P-51D kit, which provides fine raised instrument detail in the dials. A touch of dry-brushing or rubbing with a Prismacolor pencil brings it out beautifully.

many passes as it takes to score a good deep line, then flip the piece and repeat the operation on the other side. Next, bend the stabilizer/elevator to fatigue the scored styrene until things begin to separate. In the case of this P-51, a complication is the little "ear" at the outboard end of the elevator, **14**.

Your first option is to just scribe a straight line through it. Then you will have to carve the ear out of the stabilizer and reattach it to the elevator or replace it with styrene stock. Option two is to scribe around the ear and try to remove it attached to the elevator.

Let's try No. 2. This means careful deep scribing around the ear and employing a bending/twisting/fatiguing operation to separate it from the stabilizer. The only way to do that is to grip that little ear itself with stout tweezers or needle-nose pliers and wiggle it until it comes loose. Fortunately, the Tamiya plastic is soft and waxy, and with sufficient scribing, the ear comes out with little damage, **15**. A raggedy edge is inevitable, and once again some gap-filling super glue applied to the rough spots and sanded smooth takes care of this.

This bit of surgery will require cleaning up the line between the stabilizer and elevator so that when they come together again, the resulting joint is tight. It's also a good idea to mark or keep the left and right stabilizer/elevator pairs together since they will never be as interchangeable as you might like them to be. Rub a round rat-tail file on the trailing edge of the stabilizers get a slightly concave surface, **16**. Superglue a thin strip of sheet styrene to the leading edge of the elevators, sanding and shaping it to a round edge to fit the corresponding concave section of stabilizer, **17**, **18**. With the stabilizer pieces glued in place to the fuselage and level, it's time to add the elevators. You want just a slight droop—the important thing is that they both be at the same angle. Careful eyeballing ensures this. Admittedly this is a lot of effort for a small change in appearance, but it's accurate modeling.

Landing lights. The P-51B was one of those airplanes with a landing light in the leading edge of the wing. The standard way kit manufacturers have handled this is to provide a separate clear part to fill a cutout in the wing leading edge. If you ever get the chance to see a real airplane with a light like this, you will notice how the thin Plexiglas cover blends perfectly into the skin of the wing. That is not the effect a little piece of clear styrene in a cutout (even if the fit is good) provides.

To get a really smooth transition from skin to clear on the model, wrap a piece of clear tape over the opening, **19**. In the case of this Mustang, panel lines closely surround the light opening itself. With a fresh No. 11 blade, trim the tape at these recessed lines.

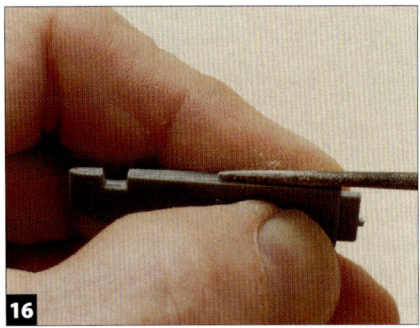

16 A round file cleans up the trailing edge of the stabilizer. Fingers act as a guide to keep the file from slipping and damaging any other part of the stabilizer.

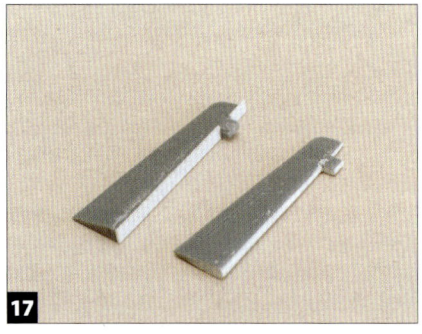

17 On the left is an elevator with sheet stock superglued in place. At right, the other elevator with the same sheet stock has been trimmed and rounded to fit the corresponding stabilizer.

18 The relaxed elevator is in place. Note the white styrene shims that were added to fill and square off edges damaged in the removal process.

19 Clear tape is wrapped around the leading edge of the wing over the landing light aperture. Using the panel lines as a guide, it's trimmed and the excess is pulled away.

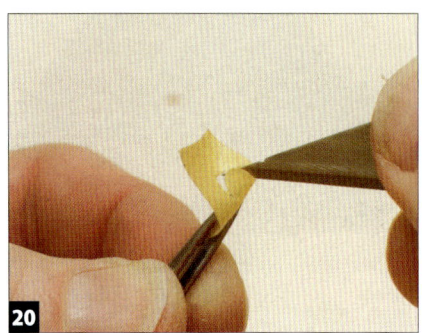

20 The kit clear piece is covered with masking tape, which is then trimmed to fit.

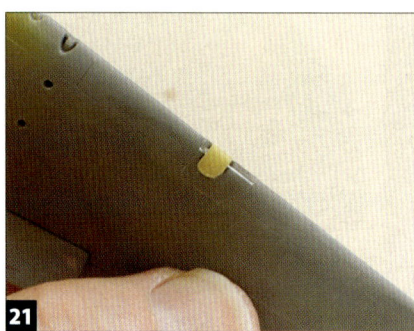

21 That tape is then used to mask off the aperture before painting.

The kit-supplied lens piece then becomes a template for a masking tape cover for the actual opening in the wing leading edge, **20**. When it comes time to paint the model, wrap masking tape over the part and trim the excess with that same fresh blade. If the remaining tape on the clear part is cut cleanly, place that tape on the clear tape matching the opening in the wing, **21**.

If it isn't a clean cut, you now have a good size and shape guide for cutting another piece of masking tape with the help of a hard-edged scribing template. Once that clear tape is on, it will be important to remember it's there. Gripping the left wing by the leading and trailing edges can mean putting a thumb on that cutout, which will likely dent the thin tape. In that case, your choices are to live with the dent (not possible for me) or starting the landing light exercise over again.

Will that flimsy piece of tape hold forever? Under a good coat of paint it's unlikely it will ever lift. If you're nervous about the permanence, a fraction of a dab of super glue on the corners should remove worries.

Paint and windows. The paint job is straightforward, with gloss white applied as an undercoat for the red areas and forming the basis of the invasion stripes. (About invasion stripes: I put them on

22 A minor modification that can be made to all kit gunsights is to replace the angled glass with a piece of thin acetate. This appears much closer to scale and is much clearer than the clear styrene of the original molded piece.

before the main paint job because it's easier to protect them with masking tape than it is to cover the entire painted model to protect it from invasion stripe overspray.)

When you have the stripes and red nose complete, let's deal with the canopy. Because it will be in the closed posi-

BUILD A BETTER LANDING LIGHT

Wrapping tape over wing leading edges can be impractical in smaller scales. However the smaller the scale, the less realistic a kit clear piece stuck in a wing looks. Here's another method to consider.

Cut a section from a thick piece of clear plastic, such as a toothbrush handle. Measure the width of the opening in the wing that needs to be filled and trim the clear stock to match it. It's easier to trim the piece by sticking it to masking tape.

Drill a hole into the backside of the clear piece to represent the actual light. Paint the back of the clear piece black to suggest depth behind it and put a drop of silver paint into the hole that's been drilled.

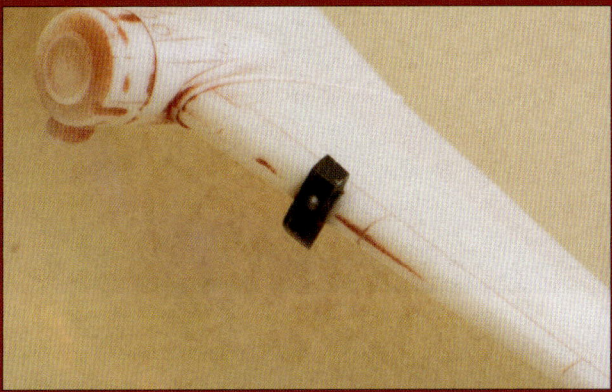

Secure the piece into the wing leading edge with super glue, making sure the silver dot is centered vertically.

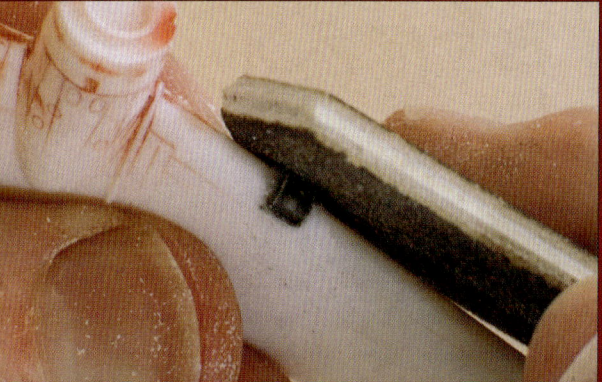

Grind the piece down to match the contour of the wing with a coarse sanding stick. Continue with medium and fine grits as the lens begins to match the wing. Since most of the pressure on the clear part is against the wing, the super glue should hold, even with the black paint layer. If it comes loose during this process, simply clean the back of it and reinstall.

After polishing, you should have something clear and shiny that looks like it was molded as part of the wing.

tion Tamiya provides, wait until the last moment to install it in order to prevent (as much as possible) dust or other small foreign objects appearing—forever sealed—behind the glass. A few blasts of compressed air into the cockpit should clear such material out of the fuselage.

Before attaching the canopy, replace the reflector glass on the gunsight with a more-to-scale piece of thin acetate, **22**. Cut a narrow strip the width of the base of the gunsight, then bend it evenly in half and trim to the correct length. Attach it with a small dab of white glue. The only trick is to position it so that the inside surface of the windscreen doesn't interfere with it. That will determine the angle of the reflector piece.

Finally, the canopy clicks into place, though not perfectly. A smear of putty on the end of a toothpick will fill any gaps, and you can sand it smooth.

The color scheme for the QP-J is the standard Army Olive Drab 41 over Neutral Gray 43. I recommend that you not get too wrapped up in questions of color correctness. There are so many variables when it comes to model color—including prototype standards, light and field conditions, aging, usage, and even the effects of scale on paint appearance—that it's best to learn to be satisfied with what looks reasonable to you based on your research into the craft and conditions you're trying to model.

Since I'm an "enamel guy," I mixed a shade of olive that closely matches 34201 on my Federal Standard 695 color catalog and a medium gray that matches 36463 with Extracolor gloss enamels.

The decals, from the old Superscale sheet 48-139, represent Howard "Deacon" Hively's "The Deacon." This plane was lost with a different pilot the day after D-Day. So when it came to "weathering" the model, the invasion stripes were left alone, as on the real airplane, they simply weren't around long enough to get too dirty.

1/48 scale RF-84 THUNDERFLASH

Scribing surface details

6

Engraved or recessed surface detail has been the standard among the major kit manufacturers since the late 1980s. Before then, raised lines and/or row upon row of rivet heads could be expected to cover their models' exteriors. Raised detail was by far the less expensive route for kit manufacturers when producing steel molds. Advances in mold-cutting technology would eventually lower the cost of recessed detail, which has been terrific news to the model airplane community. But more than 30 years of molds for raised-line plastic kits are still out there, most never to be replaced with 21st century molds.

Replacing raised details with scribed panel lines can greatly improve the appearance of a kit, such as this Thunderflash by Heller.

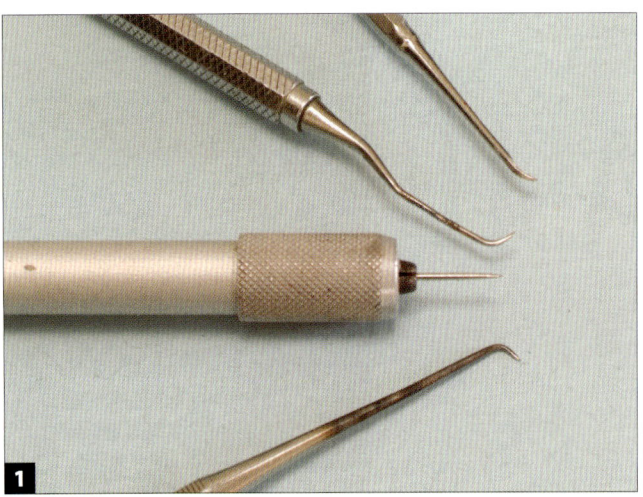

1 Any of these scribing tools will accomplish the job of scratching plastic. The needle in the pin vise is my preferred tool. Heating for some long-forgotten operation has discolored the scriber on the bottom.

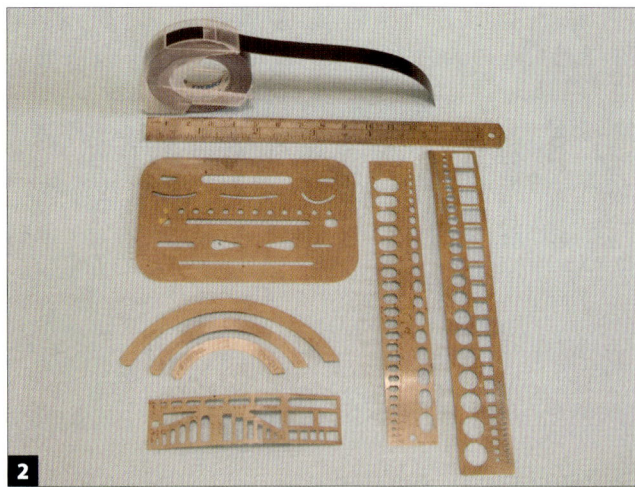

2 All of these items can serve as scribing guides in different circumstances. From top to bottom are Dymo labeling tape, a steel 6-inch rule, a drafting erasing shield, and a collection of modeling-specific scribing templates.

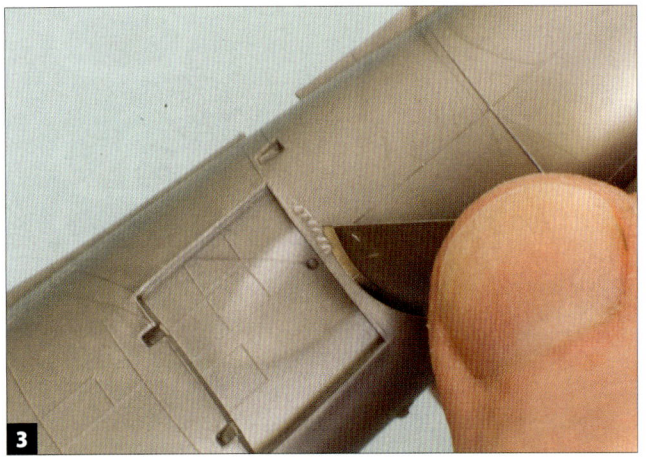

3 A raised panel line comes off in a curl with the help of a curved knife blade.

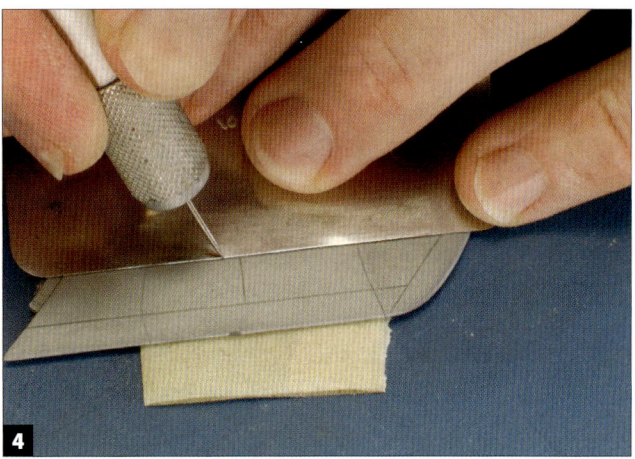

4 This is as easy as it gets. With a flat piece taped in place, the only thing that should move is the needle against the aluminum guide.

Successful scribing. So what's wrong with raised-line detail? On the surface (no pun intended), nothing. Nothing, that is, as long as the kit clicks together perfectly with no filling or sanding, and those raised lines align with each other on opposing parts. But what if the fit is poor and the color scheme you want to put on the finished model is glossy or natural metal? Since gloss and the shine of natural metal will highlight the surface detail, good or bad, that can be a recipe for unhappiness. If this is the scenario I'm faced with, I will sand the model smooth and replace the surface detail myself. Rescribing a plastic kit can be tedious and time-consuming, but the results will be extremely satisfying.

You can pick from a number of dedicated scribing tools, **1**. All are some version of a tool with a V-shaped point on a long handle. They should remove plastic in a nice swarf curl and leave a sharp, smooth line. They work well on flat surfaces, but significant portions of any airplane's exterior are not flat. More often than not I keep going back to a simple sewing needle chucked in a pin vise.

A scratching tool is half the requirement. The other essential is some kind of guide, **2**. For straight lines on a flat surface, any straight edge will do. My favorite is a simple drafting eraser shield. It provides lots of area to press down in order to hold the piece still. However, there will also be specific shapes and curved lines on curved surfaces where a straight-line template won't be of much use. Thankfully, aftermarket-modeling firms have produced some handy templates for these situations. I'd suggest obtaining some.

My first step in a scribing operation is to sand smooth the parts to be scribed. If the model is molded in silver, a visible "memory" of the removed surface detail will remain as a guide. (If the plastic is not silver, this visible memory will still be there, though harder to see.) In the case of this particular demo subject, where the end product will wear a silver finish, I don't want to use coarse paper, nor do I want to spend hours wet-sanding pieces smooth with extra-fine-grit paper. Therefore, I start the procedure by shaving off as much raised detail as possible with a dull curved No. 10 hobby knife blade before sanding anything, **3**. (A fresh blade will want to dip into the surface, leaving divots.) With a natural metal finish as the end goal, I didn't sand this one with anything coarser than 800-grit paper.

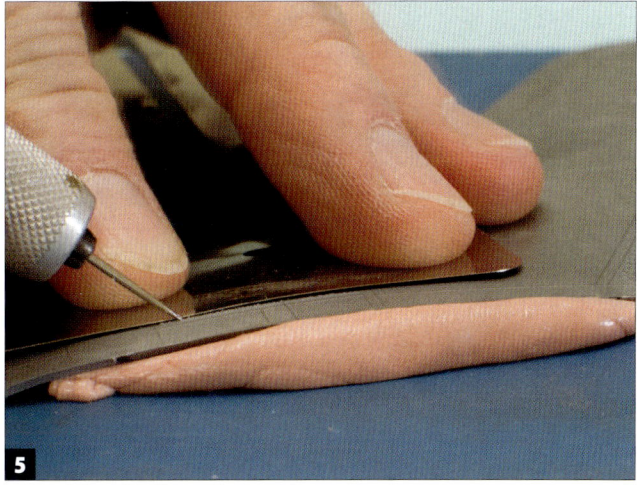
5 The pink ooze is Silly Putty. It does a nice job of supporting the slightly bulged shape of the air intake under the pressure of a pressed-on template and a scribing needle.

6 Longer lines needing to be scribed benefit from a straightedge more robust than an aluminum eraser shield. A steel rule such as this will not flex as the needle is pressed against it.

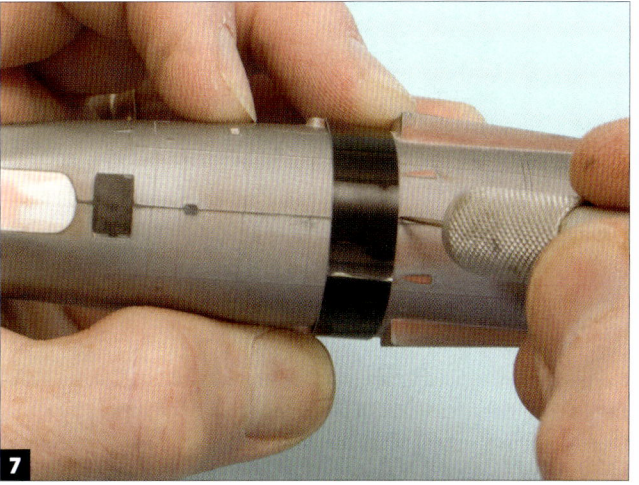
7 I see four mismatched panel lines meeting each other here. This is a common aggravation of raised-line detail. With Dymo tape stuck to the belly as the guide, the left or right side is chosen as correct, and the line is drawn across the center line accordingly.

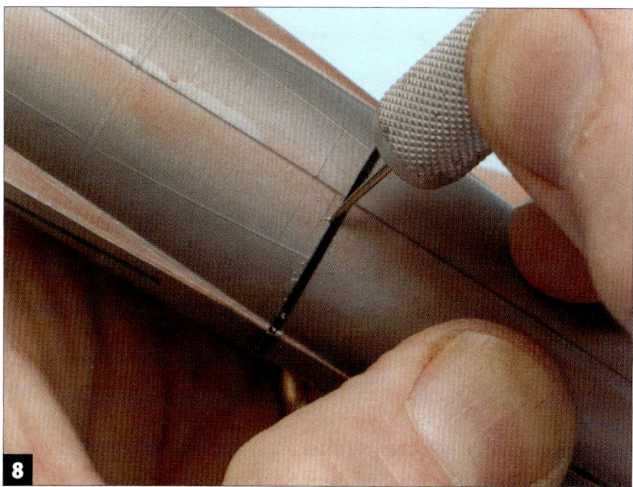
8 Scribing a line on either side of firmly affixed Dymo tape should ensure a pair of closely spaced parallel lines. If you can get the same result freehand, you have a rare talent.

Freeze! I have found that the most important ingredient in rescribing plastic is making sure the part being worked on does not move. With pressure against a smooth part from both a pressed-on template and the scriber itself, it's extremely easy for something to slip if the part isn't anchored. Starting with the easiest piece, a nice flat stabilizer, I stick it to a cutting pad with a loop of masking tape, **4**. It will not move unless the cutting pad moves. I line up my eraser shield on one of the visible lines and score it twice with my needle. Your pressure and my pressure on a needle may not be the same. There's no "correct" depth of the line you scratch, and the look you like will be gained with experience. (Personally, I do not want deep surface grooves to dominate my finished model.)

Also, there is no rule that says you must scribe back into the model every single line that was originally there. Sometimes less is more, and sometimes those lines provided by the kit manufacturer are not correct.

A minor complication with this Thunderflash is the bulge of an air intake in the wing. Modeling clay or Silly Putty keeps the contoured section of the wing from being pressed flat while lines are scribed, **5**. Maintaining the proper shape of the parts while lines are put in will ensure straightness of those lines.

This contour also complicates a long line running the length of the wing, **6**. The temptation is to scribe this in its entirety with one long scratch. It's safer to break it in two: Scribe the flat outboard half of the wing as one line and the slightly contoured part of the wing as another. (I know because this one bit me.) A perpendicular panel line at the halfway point of the long one makes a perfect beginning and end point for two joined lines. I also recommend a stiffer straightedge for scribing long lines. The thin aluminum eraser shield may flex, and your line will acquire an unwanted dip.

The majority of the lines scribed into the wings are straight and easy. There are a few, however, that meet each other on the top and bottom of the air intake, **7**. These I leave alone until the wings are

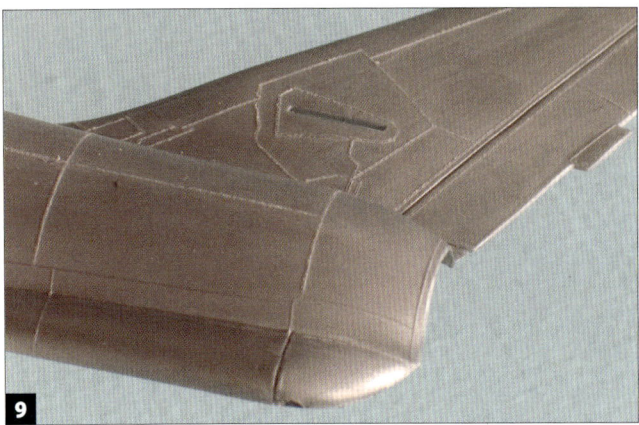
9 Scribing with a simple sewing needle will result in fine ridged grooves in the plastic. You'll want to remove the ridges.

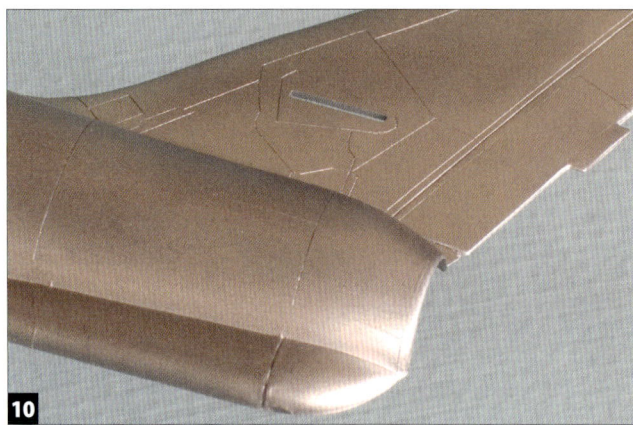
10 After a thorough wet sanding with extra-fine-grit sandpaper, a toothbrush scrubbing, and a light coat of silver paint, those ridged grooves become beautiful scribed panel lines.

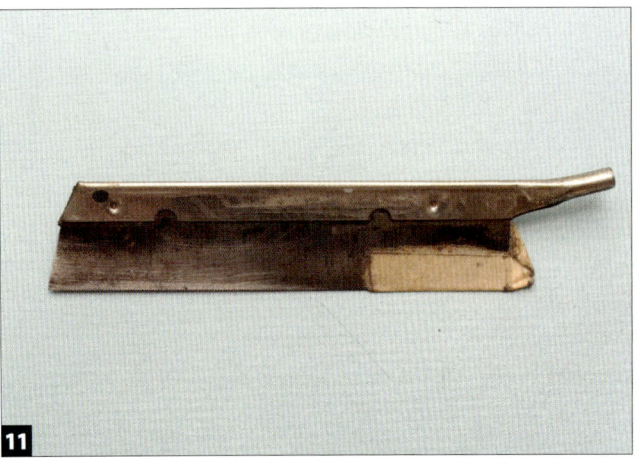
11 Sometimes a customized tool is the only way to go: This razor saw has been modified to become a panel line cutter and clearer. The set has been filed out of the teeth and the front end trimmed to get into tight spots. The masking tape prevents the back teeth from digging into the modeler's hand.

12 Only about the first 10 teeth do all the work of digging a line. Once the line is established, this tool can be easily used to clear out sanding residue or super glue filler.

assembled, ensuring that when I scribe them, there will be no top-to-bottom mismatches.

The fuselage is the fun part. I scribe as many lines as possible (the short, straight ones primarily) on the fuselage halves before they are glued together. It is much easier to work on an anchored fuselage half on its flat side than when it's attached to its opposite round side. As with the wing intakes, I leave alone any lines that will meet on the spine and belly.

It's with these lines that label tape (like Dymo brand) proves its worth. On the top and bottom of the fuselage, panel lines start to curve significantly. Happily the label tape can be cut to match curved "straight" lines, if need be. It can also be cut into thin strips, **8**, which are extremely useful for close parallel lines. Keep in mind that there will be less sticky surface touching the model, so be careful about the pressure of the scriber against tape. Careful scribing on both sides of the tape will provide perfect, even spacing of a pair of lines—ordinarily a tricky thing. Also be aware that the stickiness of the Dymo tape doesn't last long with repeated use. After four lines, it's probably a good idea to cut another piece.

In a perfect world (or with a perfect tool), scratching the lines would be the end of it. Unfortunately, when the line is scratched in the plastic, you actually create a ridged groove. That ridge needs to be removed, **9**.

To some degree, you can shave the ridges just as you shaved the original raised detail with the curved knife blade. The blade can do the heavy lifting, but with or without it, more wet-sanding will be needed. This sanding process will fill the scribed lines with the wet dust you sand off the surface. A good scrubbing with soapy water and a toothbrush should clear out most of the micro-debris, **10**. Stubborn stuff that remains can be cleared out with a sharpened toothpick or—very carefully—with the same needle that put the line there in the first place.

A valuable tool I made decades ago is good for scribing lines on curved surfaces, **11**. It is a hobby razor saw with the sides of the teeth filed smooth, removing their "set." I also trimmed the top of the blade back to allow the front teeth to get into tight spots.

The saw is held in the same manner as you would a saw-shaped pencil. I align the teeth with a vertical panel line and as much as I push the saw, I rotate the fuselage into the teeth. This results in a narrow line being cut into the plastic, **12**. With slow, steady pressure and concentration on what you're doing, no templates are needed.

Unfortunately, since the mid-1970s when I customized this particular saw blade, the brand of saw I use has slightly increased the width of its most-comparable razor saw. This makes a noticeable difference in the width of the line and the tool's performance. Mine has 40 teeth per inch (I counted them) and they face forward. Exactly duplicating my tool may be tricky, but is certainly possible.

Damage control. Inevitably there will be slips. Damage control is simply the application of super glue filler into any scratches where the scriber or saw went too far or wandered off course. When the glue cures, sand the excess off the surface. You might get away with fixing a bad line with body putty rather than super glue if the model will wear a flat finish. However, under a silver finish like this Thunderflash will wear, that putty-filled line will probably show as a slight depression.

Sometimes there are situations where an alternative to scribing certain lines may be in order. On this Thunderflash, such cases are the landing-gear doors and speed brakes. Typically, these parts didn't fit well in the closed position. Significant super glue filler was needed to plug gaps, **13**. Scribing through filler of any type risks chipping or cracking it. Also, the shape of the landing-gear doors is irregular, and the two little ears on the speed brakes would be a tedious job to scribe exactly. My solution is to mask off these areas and give them a few good coats of primer. The idea is to apply enough primer so that after you wet-sand the top layer smooth, you're left with a slightly raised panel on top of the chosen locations. Under a coat of paint, they will appear as separate, distinct panels.

Before bringing the sub-assemblies together, a light dusting coat of flat silver is applied, **14**, **15**. This instantly reveals scribing mistakes, scratches, and other surface flaws. Any problem areas are dealt with, and all the big pieces are brought together.

(continued on page 54)

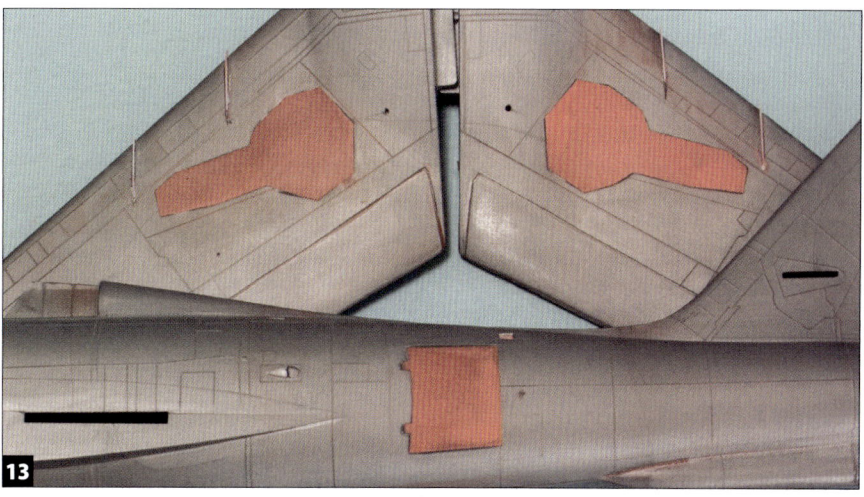
Underneath a coat of paint, these primered panels will be raised just enough to set them apart from the surrounding airplane skin.

The model is nearly ready for the silver treatment. All areas not to be silver have been painted and masked. The canopy is masked, and exposed frames have been painted black. This way they'll be black on the inside and silver on the outside.

With the base coat applied, various panels are masked with sections of self-adhering notes to be repainted with a slightly different mix of silver. Note that only flat surfaces are treated here. A paintbrush inserted into the engine exhaust serves as an excellent handle for the model with its easily damaged paint job.

RUBBER MOLDS AND RESIN CASTING

Casting your own parts in resin can be the answer for copying and mass-producing items that either can't be purchased or would be prohibitively expensive if purchased in the quantities needed. However, if cost is a factor, keep in mind that rubber and casting resin aren't exactly cheap.

The simplest mold is the one-piece open-face type. To make one, first I enclose the part I want to copy in an open box made of toy bricks (like Lego brand). With Lego pieces, you can create any size rectangular enclosure you could ever need. The height of the enclosure should exceed the height of the piece to be copied by at least an inch. This is to accommodate the rubber that will become the bottom of the mold. Likewise, be sure the walls of your mold box aren't right up against the edges of your master part. You want enough rubber surrounding the part to ensure the mold is solid and its walls aren't thin.

It's a good idea to attach the master part to the floor of your enclosure with a dab of white glue or any other temporary fixative. This will prevent the part from shifting under the weight of the rubber piling on it while it's poured. With the part centered in the Lego box, the pouring can begin.

There are many different RTV (Room Temperature Vulcanizing—hardens at room temperature) rubbers to choose from depending on the strength of mold needed or the type of casting material eventually to be poured into it. For this simple application, I like Micro Mark's ONE-to-ONE Rapid. It's a simple one-to-one mix ratio (not all rubbers are). With equal parts measured into a container, it's stirred with a stick until the blue and white blend into light blue, and it cures in a few hours. It will cure faster if placed near a heat source. When touching it leaves no fingerprints, the rubber is cured. At that point, the Lego box is pulled apart, the master is removed from your fresh chunk of rubber, and you're ready to start casting.

Like rubber choices, there are many different casting resins to choose from. I prefer Alumilite. Like the rubber, it has a one-to-one mixing ratio. Unlike the rubber, Alumilite will cure very quickly—in just a few minutes. Again, equal amounts of the clear and amber portions are mixed together, and in the process, the material goes from cloudy to clear. When clear, it's thoroughly mixed and ready to pour. Within a minute, it will thicken noticeably, and within five minutes, it will transform from a clear liquid to a tan solid. Like most resins, Alumilite generates heat as it cures. A concentrated amount will produce a surprising amount of heat. Also like the rubber mold material, the warmer the surroundings, the faster it reacts. The part should be fully cured and cool in about 10 minutes. Pop it out of the mold, and make another and another.

When it comes to copying kit parts, be aware there are copyright laws. You're not likely to get into trouble making copied parts for personal use, but if you try to sell them, the creators of the original piece could have a case against you.

The object I want to copy is seen here enclosed by a wall of Lego blocks. Attached to the bottom surface of the object is a thin sheet of styrene. This will give a raised edge to the top of the cast piece that will be sanded off to ensure a sharp bottom edge to the piece. Make sure your master is as perfect as you can make it. Any problem with it will be repeated over and over. The red-brown residue seen on top of some of the blocks is a previously used rubber.

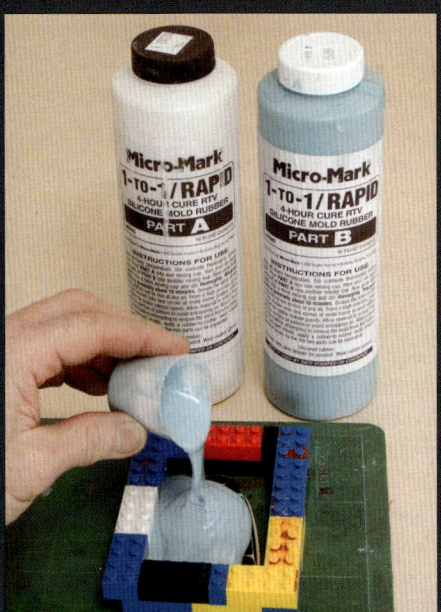

With the master fixed in place, the rubber is poured over the top of it. Good quality rubber material will flow into every crack and crevice of the master part without any help.

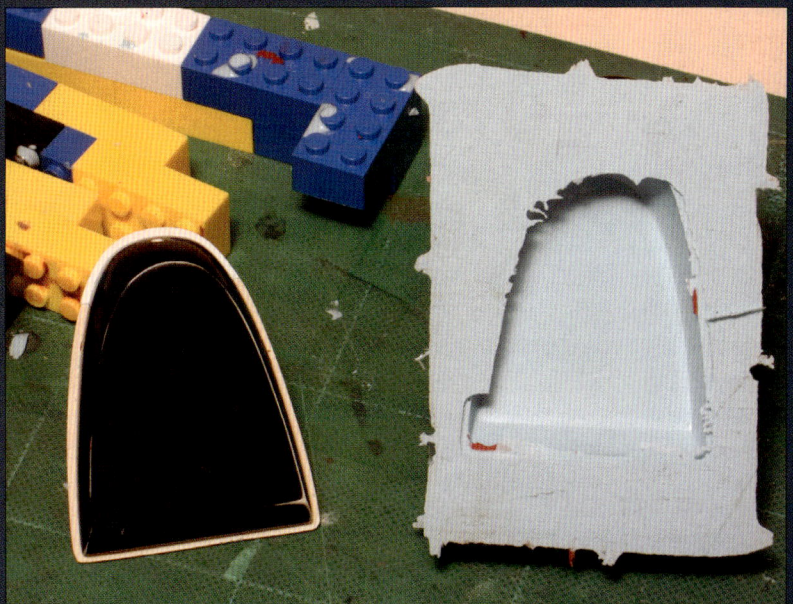

After the rubber cures, the Lego box is dismantled and the master object removed. The ragged edge on the mold is the result of the rubber seeping under the master. As stated previously, the rubber will find its way into any space possible. This flash is easily trimmed away with a razor blade.

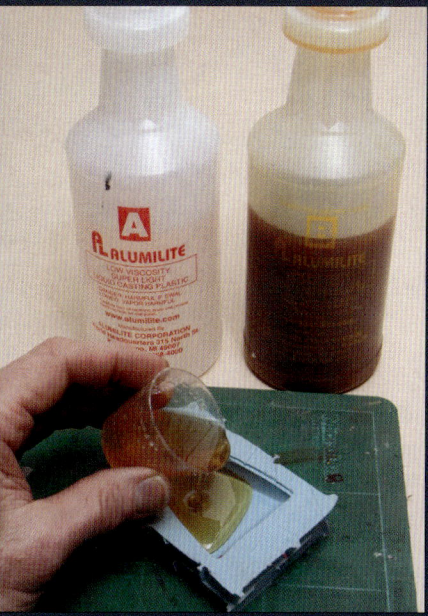

With the mold cleaned up, it can be used immediately. Here the Alumilite resin is poured into the cavity. Pour enough to just begin to fill the edge made by the lip that was added to the master part.

The object poured here will eventually form the base of stands for a collection of models in flight. A common base shape was a requirement here. It's far easier and faster to cast multiples of this sort of thing than to fashion each one individually.

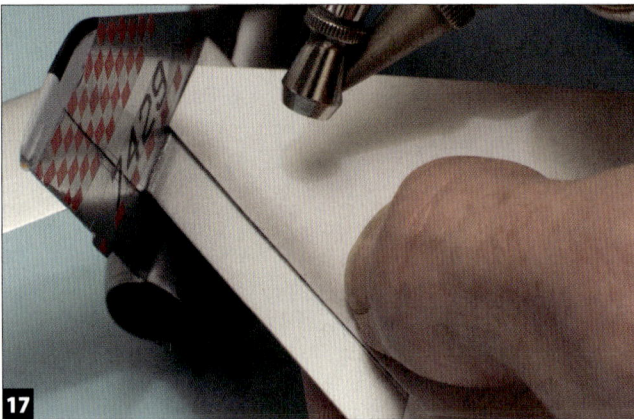

16 Parafilm-M works well for masking the trickier curved fuselage surfaces. Here a panel has been masked and carefully trimmed at the panel lines. Tweezers remove the trimmed film.

17 Airbrushing a very thin mix of what amounts to darkened thinner on one side of a panel line can effectively change the tone of the chosen panel. This is commonly called "post-shading."

(continued from page 51)

Au naturel. Natural metal is by far the most challenging airplane finish to model. My method is to first paint and then mask off anything not to be silver, such as wing walks, anti-glare panel, and trim colors. The model then gets a thorough coating of decanted Tamiya TS-30 spray can silver. This is a mild lacquer that dries very quickly. It can also be sanded when fully cured, allowing further surface fixes to the model if needed. Once I'm happy with this complete silver base coat, I spray several coats of Alclad II chrome over the model to shine it up.

With this final base coat of silver laid down, various skin panels are masked off for further painting. The trick is finding a masking medium tacky enough to stay in place, but not so sticky that it pulls up paint when removed (for which silver paints are notorious). For flat surfaces such as wings and stabilizers, my choice is self-sticking paper like Post-It brand notes.

These sticky bits of paper can be laid right on the panel line. They're easily trimmed with scissors to fit into corners and tight spots. Cut up an index card or ordinary business cards to make a rigid straightedge to paint against or to provide temporary masking with your free hand while airbrushing with the other. Unfortunately, self-sticking notes are nearly hopeless for curved surfaces.

On fuselages or engine nacelles, I use Parafilm M, **16**. This is a stretchable, waxy tape that clings to smooth surfaces when pressed against them. After a short piece of Parafilm is cut from the roll, the backing sheet is lifted and the film itself is stretched to about triple its original length. The stretched film is then pressed in place overlapping the panel you wish to paint. A moistened cotton swab will press it into corners like wing roots, and there it will stay—not on the swab. Because of its pliability, it's nearly impossible to pre-cut the stuff. This means it must be trimmed once its been pressed on the model.

In the case of my Thunderflash, the scribed panel lines will guide the point of a fresh knife blade, and with focus and a deft touch I can ride those lines with my blade. Make no mistake: This is a game of chicken. No cut except right on that panel line is allowed.

Panelizing: Easy does it. Another area where less is more is "panelizing" with paint. In most cases, the differences in tones and textures on real airplane aluminum skin panels is subtle. For me, this kind of subtlety translates to a model best when the variations of tone are mostly revealed depending on how the light hits the surfaces. Although we have a wide range of metallic paint shades to choose from, a realistic look can be had simply by overspraying select panels of the final base coat with gloss or flat coats. Areas around exhausts are usually noticeably darker than the rest of the airplane's skin, and adding a little bit of black to the basic silver makes that difference on a model. With this masking and re-painting complete, all the masking tape comes off and the decals go on.

One quite unrealistic effect of this silver finish is reflection off of corners and recessed lines where, if anything, there should be shadows. In the case of the spaces between aileron, flap, and wing and rudder and stabilizer, I lay a line of artist's black oil paint down and blend it smoothly and evenly with a soft brush. Any black that gets out of the recess is easily whisked off the surface with a cotton swab. I apply the same basic technique where the wing fences meet the wing. Putting a dark sludge wash into all the panel lines is something to consider, but on this one, I left them alone. (It's been enough work as it is.)

Post shading. To add more visual interest, I employ a technique known as post-shading, **17**. This is simply airbrushing a very thin mix of black (or dark brown or dark blue if you prefer) against a mask laid down on a panel line. The unmasked edge receives a soft dark edge. I also post-shade evenly spaced lines on large panels just to "busy" them up a bit. Subtlety is the key. The temptation is to keep spraying until you clearly see the effect, but by then it will be heavier than needed. This post-shading should enhance, not dominate. Let the angle of the light reveal this work.

The display stand is essentially an angled box made of .060" styrene sheet. It is filled with old sprue and resin junk and anything else to provide some ballast and solidity. The clear stem is ¼" extruded acrylic rod, oven-heated to about 375 degrees for a couple minutes and bent to a nice curve. I drew a bull-shaped template and laid it on a yellow disc of decal stock to create the squadron emblem. The lettering is computer generated and printed on clear decal film.

BASIC BASE

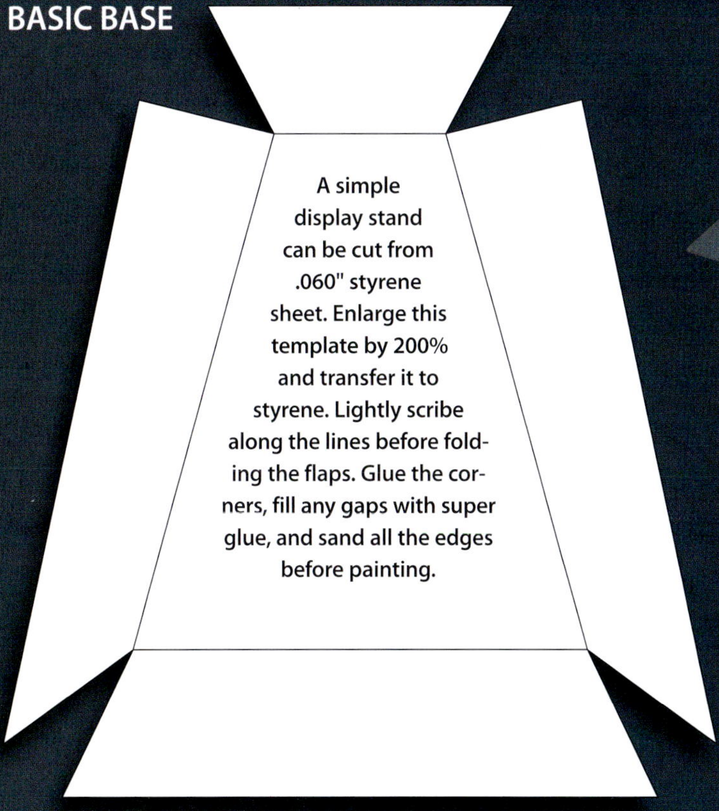

A simple display stand can be cut from .060" styrene sheet. Enlarge this template by 200% and transfer it to styrene. Lightly scribe along the lines before folding the flaps. Glue the corners, fill any gaps with super glue, and sand all the edges before painting.

The mounting post is ¼"-diameter clear extruded acrylic rod, heated and bent to shape. Drill a hole in the rear of the box to mount the rod. The box needs to be weighted with ballast to provide a stable base for the plane. Decal lettering and a clear coat finish the base.

Adding aftermarket details

Resin, photoetched, and other aftermarket detail parts can greatly enhance the realism and finished appearance of a model.

When I began building plastic airplanes, any details modelers wanted to add to their kits came from their own ingenuity. "Aftermarket accessories" meant sheet styrene or index card stock, toothpicks, soft wire, and anything else you might stumble across. Today, there's almost no end to the instant detail you can buy for your models.

Resin. The most popular items are resin replacement cockpits, wheels, wheel wells, gun bays, and the like, **1**. Some of these castings are so intricate and detailed they must be seen to be believed. Such detail parts need to be packaged and handled with great care. Detail is the good news.

Disappointment is a possibility when you start fitting these brittle resin pieces onto your kit. The instructions will tell you which particular kit the resin part was designed and produced to complement. You would think that means those resin parts will easily interchange with the original styrene details of the suggested kit. When they do, the resin experience is a joy. However, sometimes a significant amount of grinding, chopping, test-fitting, more grinding, and test-fitting again is the reality. Unfortunately, the more you have to handle fragile items, the greater the chances of damaging them. When this is the case, the modeler may question the value of this sometimes-pricey "instant" detail.

1
An assortment of interior items from a few different manufacturers in 1/72 and 1/48 scales are shown in this grouping.

2
A variety of photo-etched details is available. Note the difference in color of the various metals. The chrome silver parts can be stainless steel and, even when this thin, can be tricky to bend. Brass is much softer and easier to trim and manipulate.

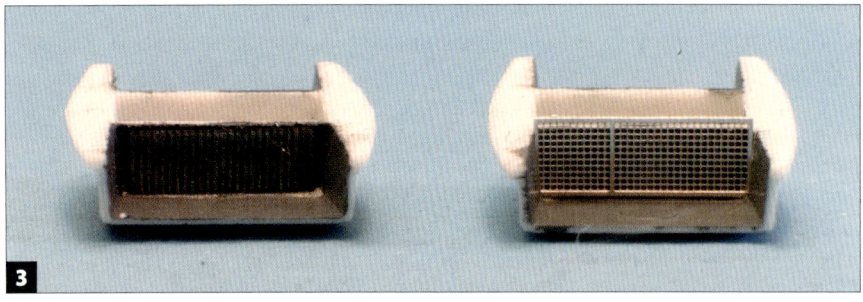

3 Each modeler has to decide where to stop with detailing. Will anyone really notice the improvement of the photoetched Spitfire radiator at right, or is the kit part, with just a little picking out of the highlights from the shadows with silver paint, good enough?

The resin material is best suited to small chunky items like wheels. Resin can—and often is—used to make very fine objects. However, such items are often too delicate to handle. Removing them intact from the shoe they are attached to can be nearly impossible. Also, the finer and thinner the object, the more prone it is to warping. For this reason it's hard for me to take something like a resin pitot tube seriously. I thank the manufacturer for their trouble and use their piece as a template to make my own.

Photo-etched metal (PE). The other major form of instant detail is the class of fine metal items produced by the photo-etching process, **2**. The obvious difference between resin detail and photo-etched is the three-dimensional quality of resin versus the two-dimensional quality of the metal. Metals from soft brass to stiff stainless steel can be photo-etched, and of course the softer the material, the easier the parts are to work with.

There may be some minor detail relief in the metal parts, but they are essentially flat. In some cases, such as instrument panels, grab handles, and levers, this is just fine. In other cases, these flat objects must be bent or rolled to form the three-dimensional items they're meant to represent.

Dedicated bending tools such as the Etch Mate, Hold & Fold, and Mr. Fold-It help when working with photo-etched parts. You may want to consider such a tool if you plan to make metalwork a significant part of your building. However, most bends are not particularly intricate, and nothing more exotic than two blades or a pair of tweezers are necessary for the occasional piece of added metal detail.

In some situations photo-etched pieces are the perfect answer to a problem, such as tiny ring sights on machine guns, fine control horns and spoked wheels on WW I subjects, or those fancy rudder pedals found in most Luftwaffe airplanes. In other situations, even though the photo-etched part is provided, using it would probably make no improvement to the finished model and would only complicate the assembly. Some modelers are just excited to add metal bits to their creations and figure the more the better. Personally, I'm not fond of working with the stuff and debate the pros and cons of any piece I consider using.

Some items you may encounter will be so tiny as to be impractical to all but a brain surgeon. Once you have the piece removed from the fret, you must grip it to place it where it belongs. Tweezers are the natural tool for this. However, gripping a tiny metal part in the points of metal tweezers may produce a tiddly-winks effect, with the part escaping the tweezers and traveling a great distance. Will it be found? Will it be worth the time spent searching for it? All you can do is be very careful. Or you may want to consider

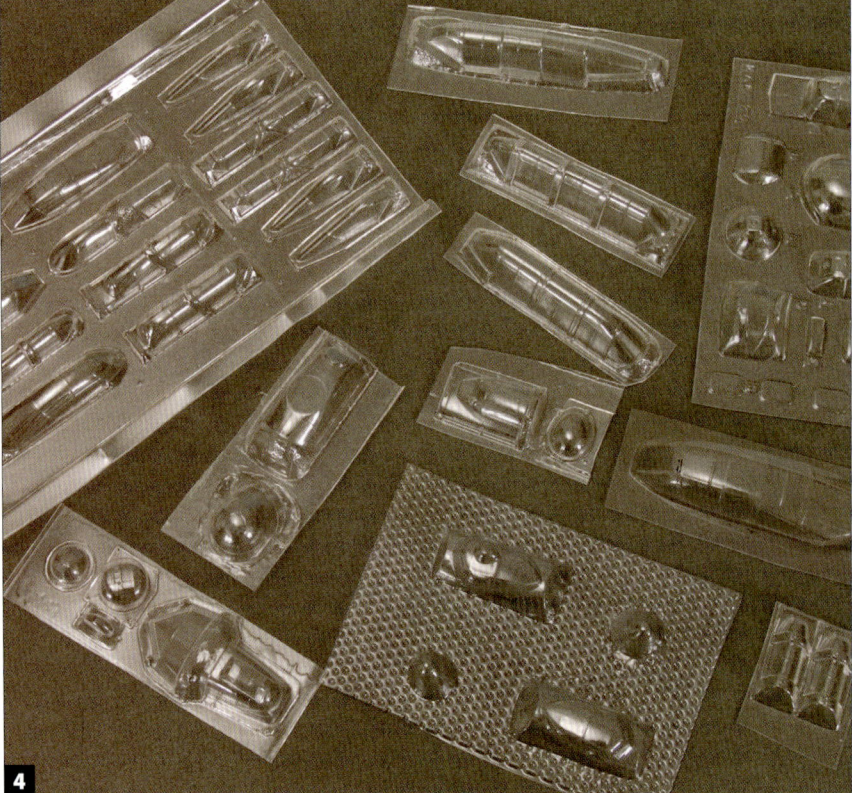

4 Here's a collection of vacuum-formed clear parts. Most are examples produced by Falcon in Australia, which has largely cornered the market on these items.

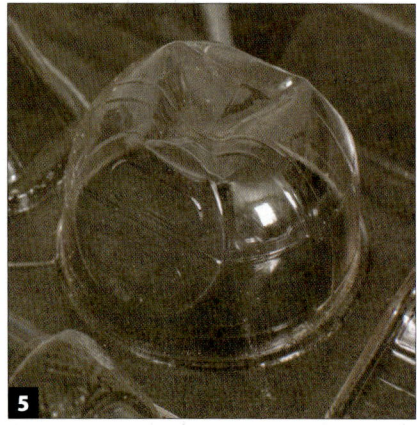

5 This is a good illustration of what can happen to a thin plastic protrusion that takes a hit. There's no fix other than replacement.

dedicating a pair of tweezers with bits of fine sandpaper superglued to the ends for handling photo-etched parts.

A project for this book illustrated to me one of the debates I mentioned above. Having four different radiator grilles (two front, two rear) for a Spitfire, and not being positive which was which, I started looking at reference photos. Even fairly close-up shots weren't much help as the area in question was always set into a box deep in shadow. Realistically, I thought, on a model it should just be dark in there. A bright chrome grille piece may look cool and might impress the penlight police at model contests, but how realistic is it? Wouldn't it tend to make the model look more like a model than a shrunken airplane? I suppose it's a judgment call. In my case, I painted the radiator kit parts flat black, dry-brushed a little silver and called it done, **3**.

Occasionally, I see queries on on-line modeling forums about the suitability of a particular photo-etched set "working" in a kit it was not designed for. By working, I assume they mean fitting. No matter the kits involved, you've got to make PE parts work. They don't jump off their fret and click into place. In the case of interior sets, most of the pieces are meant to be added on top of what's already there in styrene. Sometimes a piece is provided that is to be bent until it looks like a seat. One would like to assume a photo-etched 1/48 scale P-38 seat, for instance, would work in any 1/48 scale P-38 kit. I would say that, in general, most of any given photo-etched detail set should work in any kit of that particular subject. If for some reason it does not, this is where modeling skill comes into play. If you can't make it work, maybe you can make your own item.

Vacuum-formed canopies. A third aftermarket accessory we'll look at is the vacuum-formed canopy, **4**. The Squadron line of Crystal Clear canopies is the main source for this item. Not only do they represent the scale thickness of clear areas on airplanes better than most clear parts included in kits, they can make an open canopy much easier to achieve. If you've ever tried to show a canopy with telescoping sections like those on an SBD or an SNJ/T-6 in the full open position, you'll appreciate vacuum-formed canopies.

6

Here's the author's veteran vacuum-forming machine. A sheet of plastic is clamped and suspended over the hot plate on the left. When the plastic softens and starts to sag noticeably, the clamped plastic is swung over to the right to settle on the piece that serves as the master. The master would be parked on the opposite platform, which is perforated with dozens of tiny holes. The handle on the right side is then pumped up and down, sucking out all the air between the soft plastic and the master. If the seal between the plastic and the edges of the vacuum platform is airtight, the sheet plastic envelops the master part and is ready for trimming.

You'll also find vacuum-formed canopies in vacuum-formed kits, resin kits, and most limited-run kits that aren't injection-molded. Like everything else in this hobby, the quality will range from great to useless. Hopefully they will be protected in some way. Nothing in modeling is more fragile than a thin vac canopy, **5**, and too often, they are just tossed in the box like an ordinary part of the kit. Because of the possibility of poor quality or damage or both, there's a good chance that at some point you'll want to make your own vacuum-formed—or at least heat-stretched—canopy. In theory, making a canopy from clear sheet stock is relatively easy. The tricky part is making something that exactly matches the contours of the opening in the model it's meant to cover. After that, it's a question of how fussy you are over the actual clear and smooth quality of your piece. I've never been able to avoid minuscule bumps in my own homemade canopies, no matter what clear material I've used.

By far the handiest tool ever produced for making your own vacuum-formed parts was Mattel's little vacuum-forming machine, **6**. This item was produced as a toy in the early 1960s, but had a short commercial life as a result of kids getting burned on its heater. They still turn up on eBay from time to time. The only real flaws they possess are the relatively small size of the subject matter they can handle and the thinness of the sheet stock they can realistically work with: .030" is pushing the limit.

Detail sets. The aftermarket can be the answer to your prayers. I don't hesitate to replace a kit seat with a beautiful resin casting with the belts and buckles already in place and needing only a paint job. Be aware, however, that like the kit you're applying the aftermarket detail to, they can have accuracy and fit problems.

The fit problems are usually just annoying. (But sometimes they're ridiculous, and when the time required to get everything straight and square is factored into the project, the true value of the detail may be greatly decreased.)

Inaccuracy issues are harder to deal with. This is especially true with photo-etched detail, which is difficult if not impossible to modify. If your PE part is too big or too small, the best use it may have is serving as a rough template to create your own properly sized piece.

59

APPLYING PHOTOETCHED METAL 1/32 scale Fi.156 Storch

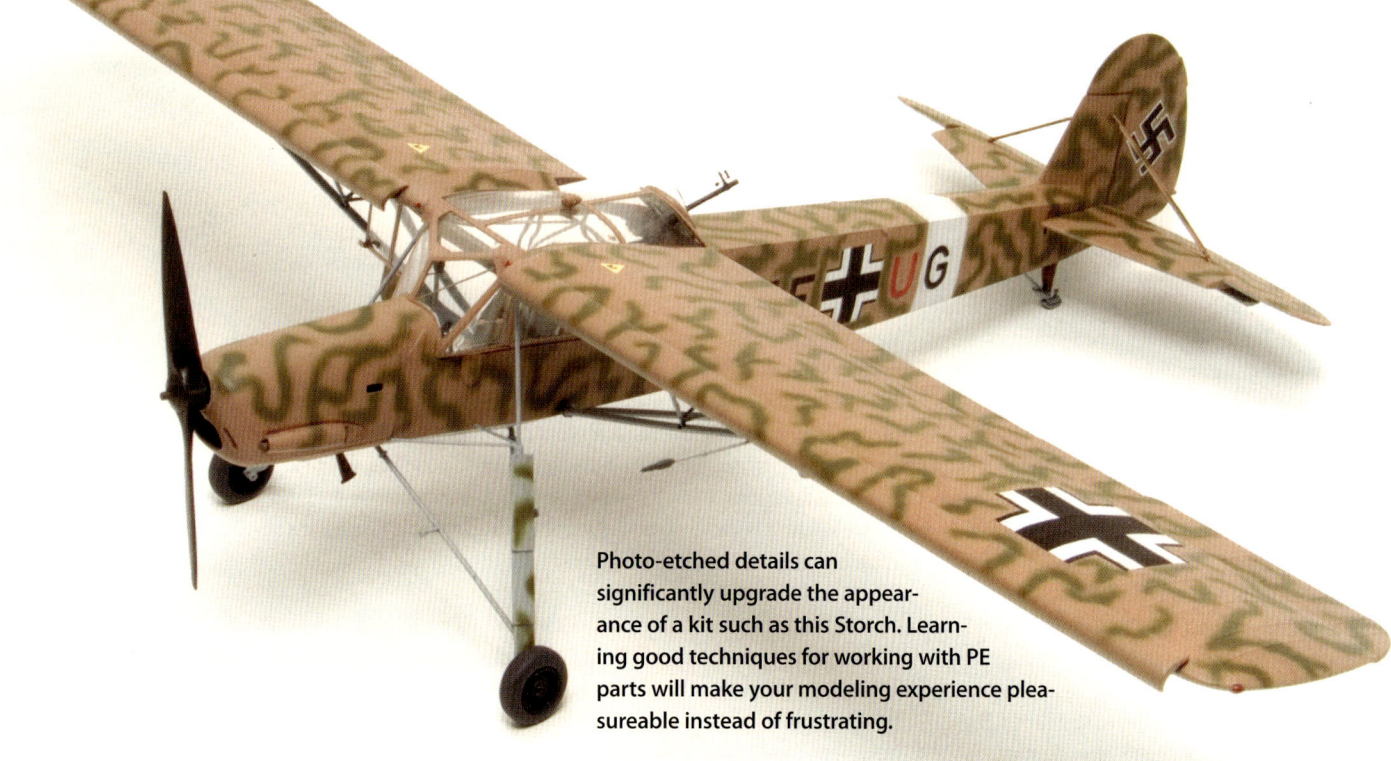

Photo-etched details can significantly upgrade the appearance of a kit such as this Storch. Learning good techniques for working with PE parts will make your modeling experience pleasureable instead of frustrating.

Upon first view, the two-fret photo-etched set is intimidating, **1**. So many tiny silver pieces! Also included is a printed film piece for the instrument panel and a couple of other separate instruments. The Eduard instructions can be a bit vague, consisting only of exploded views and symbols. Once you understand the symbol key, the project is less daunting. The first step is to establish which parts of the kit are to be replaced with metal and which are to be modified with metal.

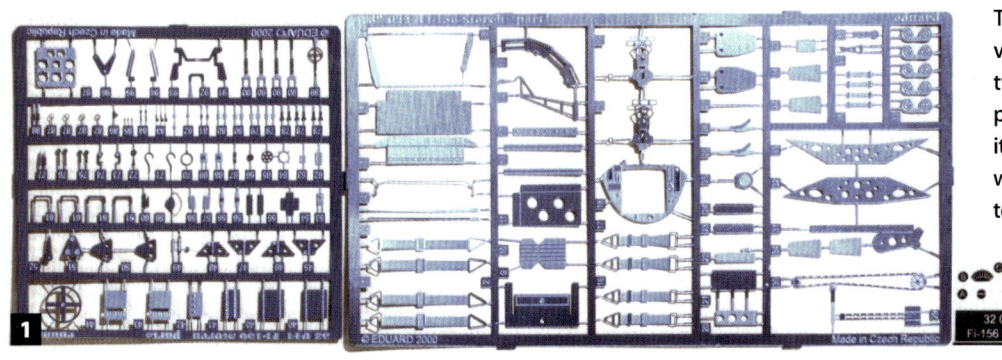

This is the starting point. This very complete set consists of two frets and almost 100 parts, plus instrument face film. Some items are easily identifiable, while others look pretty mysterious. In many cases simple bending will turn them into recognizable cockpit equipment shapes.

FINE FOLDING

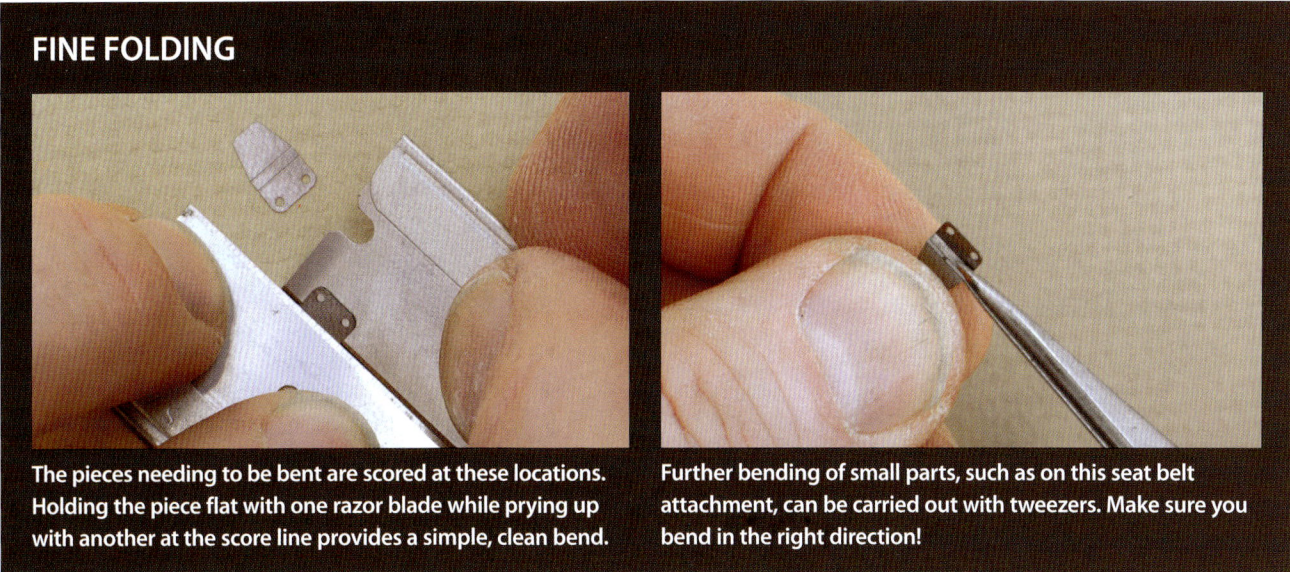

The pieces needing to be bent are scored at these locations. Holding the piece flat with one razor blade while prying up with another at the score line provides a simple, clean bend.

Further bending of small parts, such as on this seat belt attachment, can be carried out with tweezers. Make sure you bend in the right direction!

Take your seats. Let's take on the big parts first, beginning with the seats. This means assembling and attaching belts. To me, photo-etched seat belts look exactly like photo-etched seat belts and not miniature canvas or leather straps. Metal (unless it's lead foil) just doesn't bend and wrinkle realistically. I did the best I could with what was supplied and assumed painting would make a big difference in the overall appearance. In addition to belts, there are frame pieces for the seat bottoms to deal with. Bend all items as necessary and superglue them in place.

Early on, we'll be confronted with a problem that can plague projects having a lot of PE parts. Super glue does indeed connect metal to metal, but not very securely. This is one of those good news/bad news things. Since you don't have a lot of working time with the small amounts of super glue used to connect parts, often the part isn't in the exact location you want it when the glue cures. It's easy to pop the piece free, scrape off the glue, and try again. Unfortunately, if the part is in the exact location you want, it's just as easy to inadvertently knock the part off, and you get to try at least one more time. When the seats are assembled there will be 14 pieces of metal attached to them. "Handle with care" is not a strong enough warning.

Another significant item here is the rudder pedal assembly, **2**. Eduard supplies beautiful renditions of the standard Luftwaffe rudder pedals and the bar they need to be mounted on. The pedals themselves need to have a bend for the rounded heel piece and also require the leather strap over the pilot's foot. After removing the pedals from the fret, I placed them on a cutting mat and pressed and rolled the back end of a paintbrush over the heel piece, which caused it to curve appropriately. I did the same with the strap pieces and then threaded the longer piece through the tiny buckle opposite it. The bar itself required a few straight bends, then the pedals were glued on the stubby arms provided, **3**.

On instrument. The next challenge is the instrument panel. My plan was to take the Hasegawa kit part, remove the molded raised detail, and layer on metal and printed film. First, paint the backside of the instruments an off-white. I chose a radome tan. With the main metal panel piece attached to the now-smooth styrene piece, center the closely trimmed instrument film in the opening provided. Then carefully test-fit the separate metal gauge face piece on top of the film. When you see that it lines up perfectly, secure it in place with a minimal amount of super glue, making sure the glue placement doesn't allow it to creep onto the instrument faces when the face piece is squeezed flat on the film. You now have seven additional lever, handle, button, and data card bits to attach to the panel.

You'll encounter items that need some kind of depression for location, but you won't find the requisite holes. It is possible to superglue these pieces in place, flat surface against flat surface (which I did), but

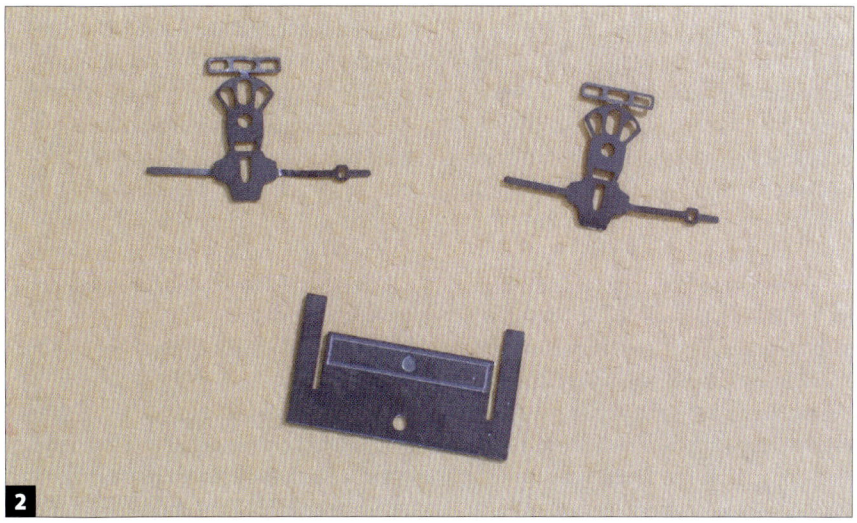

2

The rudder pedals are easy to see, but the bar on which they mount doesn't look like a bar in its raw state. After a few bends, it can be seen for what it is.

3 The rudder pedals are installed and the seats have acquired belts and metal framing underneath. The lap belts consist of five separate pieces. Though careful painting will add some life, photo-etched belts don't conform to the seat the way real canvas and leather would.

4 With components painted, the floor and seat assembly is ready to install. The instrument panel is also shown. You can't beat the effect of glass-faced instruments that Eduard's approach provides. Brown paint was dabbed and dry-brushed to suggest wood showing through worn paint on the floor.

the gluing surfaces are so tiny that fragility becomes a big issue. Also, anyone capable of working with these minute metal parts is no doubt capable of working with and making some of these parts in styrene. A styrene-to-styrene bond would be many times stronger.

Well, this is about photo-etched parts, so that's reluctantly what I used (most of them, anyway). One piece flew out of my tweezers and disappeared, and another (a mount for the external instrument light to be wrapped around the kit piece) was too tiny to be taken seriously.

Interestingly, the only photo I have showing a Storch's instrument panel and rudder pedals (in Squadron/Signal's *Storch in Action*) doesn't agree entirely with what Eduard provides, **4**.

Back-detailing dilemma. This points to another problem with the degree of detailing available today. In the absence of comprehensive references beyond Eduard's sketches, exact locations for some items are hard to determine.

In this case, a few items needed to be located on the walls of the fuselage. The throttle quadrant assembly is five pieces. I replaced two of the mounting pieces with sheet styrene and sandwiched the throttle and mixture handle between them per the instructions. The mixture handle is about twice as long as it needed to be, so I trimmed it with my manicure scissors. You can make a realistic round knob on a short piece of styrene rod using a tiny dab of five-minute epoxy for the throttle lever handle. A pair of amazingly reproduced bicycle chains are attached to definite locations on the left side of the fuselage.

Put it all together. With the cockpit floor and seats subassembly together and the cockpit walls tricked out and painted, it's time to bring it all together, **5**, **6**, **7**. The rear-most pair of cockpit support bars interferes with the belt guide pieces attached to the rear seat. Because you can't adjust the location of either the seat or these bars, it's best to remove the belt guides.

The middle support bar on the left side mashes against the petite bicycle chain, threatening to bend it. In my case, it popped the attached box on the far end loose. I reluctantly removed the rudder trim chain assembly and carefully trimmed down the tiny box at the far end so it fit closer to the cockpit wall. I permanently lost the perfect straightness of the flexible chain as a result of it being pressed by another part of the kit and my own modification.

Before the instrument panel was firmly in place, no less than five little pieces flaked off, one vanishing. Once the panel was located where it would never move again, I carefully replaced the pieces.

Because of the size of the canopy and the amount of it that will be visible to anyone caring to peer in (and why else would we be going to all this trouble?), all of the inside framing surfaces had to be masked and painted, **8**. This is no small job in these tight confines. Handle the rectangular side panels like those for the SBD canopy in the first project, with straight-cut strips of masking tape.

I robbed a canopy from a second kit to help with the non-rectangular panels. I pressed thin masking tape on some of the more irregularly shaped panels, then the raised frame lines were rubbed with the side of a pencil. Once those pencil marks were established, the tape was removed and stuck to a cutting mat where the lines served as cutting guides. This resulted in reasonably close masks, although in most cases a little tweaking was still needed.

The big cover up. With the canopy interior masked, we're at the point of no return—test-fitting this big cover, **9**. The bicycle chain that runs parallel to the forward support bar ends at the top in a sprocket that has been seriously jammed between the canopy corner and the upper-left corner of the cockpit roll cage assembly. This resulted in more unsightly bending of a long and very delicate photo-etched part. The only place that sprocket could conceivably end up is right in the outboard corner of that cage. Unfortunately, right there is where a brass landing gear strut must also pass through.

You'd think you could just trim off the sprocket and hide the ends of the chain in that corner, but without that sprocket connection on the end of the chains, it's impossible to get them straight and appropriately parallel to each other. This is a perfect example of what can happen when trying to match up such an extensive amount of photo-etched details to a kit that was planned long before the advent of these extra parts. After fussing with this part, I reluctantly scrapped it.

Included in the Eduard set are inside and outside hinges for the small triangular ventilation windows, plus opening handles. Although these are simple stick-in-place pieces, they are absurdly small to work with. I recommend you not bother with them.

The Storch comes with a nicely detailed engine and separate panels you can leave off to expose it. However, for the purposes of this project, I glued the side and bottom cowl panels together and, with the engine in place, sealed it forever behind the cowl. The fit of the cowl pieces to the front of the fuselage wasn't great. I filled the low spots with super glue, sanded the cowl smooth and rescribed the panel lines.

Cabin fever. I hope that big canopy fits pretty well for you. Mine required only a little gap-filling at the rear seams. Separate angular windows that make up the bottoms of the side bay windows were another story; all needed trimming to achieve a decent fit. I masked off the cockpit with a feeling of déjà vu, though doing the outside surfaces was easier than the inside.

Painting this model should be easiest with the wings separate from the fuselage, **10**. I wanted to finish my Storch as a desert warrior. Squadron's *Storch in Action* book (page 33) of 5F+XK provided inspiration. It appears that an irregular pattern of green had been field-applied over the desert yellow. I'd base my model on this look.

After masking off the white theater band (I looked for but did not find evidence of the white wingtip bottoms that were usually part of this African-theater marking) and the light-blue fuselage bottom, I sprayed the airframe with a light brown mix. When flat-coated, the color was a match for FS 30227.

Next come the decals. They're from the Minicraft release of the kit and were

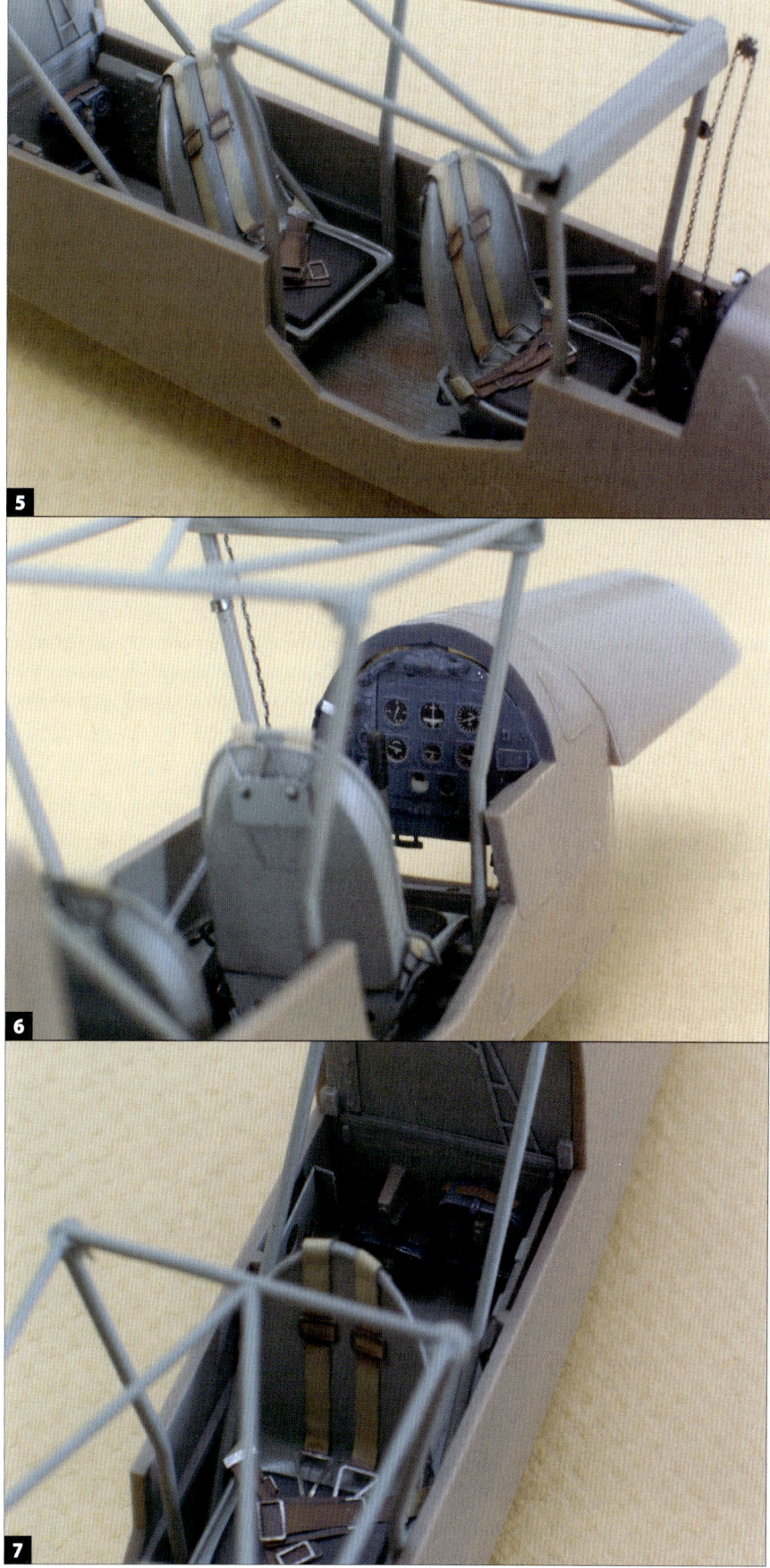

These views show the cockpit largely finished. Regrettably, the vertical flap actuator chain seen at right in photo 5 would not survive to the end of the project.

8 All interior canopy framing was masked and sprayed the standard RLM 02 gray-green primer. With the size of the canopy interior and a large open door to look into it, seeing the primer through shiny clear plastic on the inside wasn't acceptable.

9 With the canopy mounted and its exterior now masked, the model is nearly ready for a paint job. A V-strut from the fuselage to the landing gear strut proved to be a little bit short and was replaced with styrene rod.

of good quality, considering their age. Of the three marking options provided, there were no numbers in the fuselage codes. I modified S's with some slicing and repositioning to make 5's for my 5F. The other letters I chose from the kit sheet at random.

Then I mixed an olive green and wielded my airbrush as I imagined an Afrika Corps ground crewman may have—probably in a hurry and with little guidance. I worked around the insignia and code letters on the fuselage as is seen in the photo of 5F+XK. I also imagined that due to the time and effort involved with masking the complicated canopy, this minor area would have been left alone.

With the wings, I lightly sprayed random curved lines and shapes, concentrating on keeping them relatively similar in size and proximity to each other. Once I had them "sketched" so to speak, I went back and painted them seriously. Because I was literally making this up as I went along, I focused on general continuity (at the same time wondering if one guy or many guys applied green paint to the same airplane under the desert sun in 1943).

After flat-coating, I anxiously removed the canopy masking. A little paint vapor must have seeped into the cabin and deposited itself on some of the windows. Fortunately with the big side door missing and the round hole for the gun position in the back, it wasn't too difficult to reach in with some cotton swabs dipped in plastic polish and clear the windows again.

The wings fit in place with hardly a seam to be seen, which was good news. The struts found their locating holes without a problem, though the pins of the main struts at the fuselage connection had to be trimmed due to the cockpit floor being right behind the holes. The cabin door hinges fit tightly enough into their corresponding slots that gluing wasn't necessary. The MG 15 machine gun fit loosely in its mounting hole, so I set it in place and filled the hole with five-minute epoxy. This was the best ball-and-socket mount I could come up with. When the epoxy cured, I carefully attached the twin photo-etched sights and popped the gun and its window in place in the canopy. Looking good.

With the simple attachment of some remaining detail parts, careful touch-up of some of the canopy framing areas and painting of molded-on lights, I called it a wrap.

Was it worth it? Does the model benefit from the addition of photo-etched parts? Yes. After this experience of working with them, I must say that they require an enormous amount of patience. It was frustrating to have some of the metal parts interfere with the fit of plastic parts. I was left to second-guess the correctness of my kit assembly, though there didn't seem to be much room for error in plastic parts placement. I suppose this is just one of the hazards of combining parts of two completely different sources into one model. Let the builder beware.

10 For ease-of-handling purposes if nothing else, the wings were painted separately from the fuselage. Lightly suggested green blotches were applied first to get a quick balanced pattern. With the pattern established, I went back and painted the blotches in earnest. During the second round, I could tweak or expand individual blotches as needed.

COMBINING KITS AND ADDING RESIN PARTS
1/48 scale Spitfire Mk.IXe

In 1999, the Portuguese company Occidental released a much-anticipated 1/48 scale kit of a Spitfire Mk.IXe. The Occidental kit is fairly accurate in outline except for its nose, which looks slightly inflated. Let's use the Occidental kit to demonstrate how to combine parts from different kits, popularly known as "kitbashing," and for adding aftermarket resin detail.

Fortunately another company, Otaki (also released as ARII and Airfix), has produced a Spitfire Mk.VIII, and there aren't a lot of visual differences between a Mk.IX and a Mk.VIII. Most importantly for this exercise, the Otaki kit's nose contours are good. We'll use the Otaki kit from the fire wall forward to improve the lines of the Occidental kit.

Though the Occidental interior is respectable, True Details made a resin cockpit specifically for it, which we'll also use, **1**. In addition, I purchased a couple of different Eduard "Zoom" Spitfire IX photo-etched sets (not designed for the Occidental kit) to see what useful items may be found on them, **2**.

The first step to building this model is to compare the Otaki and Occidental fuselage and wings to determine if this nose-swapping plan is feasible. The initial indications look good: The respective fire walls are in the same place on both kits. Before grabbing a saw, though, some refining is in order. The exterior of the Occidental kit has a light grainy texture that needs to be smoothed. Also, the gun bay doors are supplied as separate pieces. Once they are slightly bent to fit flush in their respective wing openings, they can be solidly glued in place.

Wet-sand all the exterior surfaces smooth with 320-grit paper. (I used a worn Norton 150-grit sanding pad to accomplish the same thing as wet sanding.) Because this model will wear a flat finish, we don't have to be nearly as fussy with the surface smoothness as we were with the RF-84, but the surface grain definitely has to go! Don't bother with the horizontal stabilizers, which are decidedly thick; we'll replace them with Otaki items.

Plastic surgery. Armed with a razor saw and a miter box to ensure a straight cut, **3**, you're ready for the nose job. Even with the help of the miter box, making four perfect cuts may prove to be elusive. Nose parts are taped (not glued) together and fuselage halves are taped (not glued) together. With a good dose of liquid cement to allow for sliding and adjusting, match up the Otaki and Occidental items. Don't allow any glue into the center line, though. We're a long way away from gluing the fuselage halves together at this point. With the best compromise alignment you can manage, let the cement cure, then follow up with super glue to fill gaps. You can further reinforce the joint from the inside with a section of sheet styrene and thick super glue.

Inspection after the grafting procedure shows slight contour disagreements, which are to be expected. Otaki's cross section isn't as round as Occidental's, **4**. Also the front of the Occidental wing will now be required to match Otaki parts. Fortunately, there's nothing some sheet styrene or epoxy putty and super glue can't make right.

True Details designed its cockpit with a piece of the floor molded to each of the side walls. The idea is for these two pieces to lock together at the floor to make a one-piece insert. However, even if the parts were not warped slightly (they were in my kit), there would be problems. With a razor saw, separate the side walls from the floor sections and deal with each piece individually.

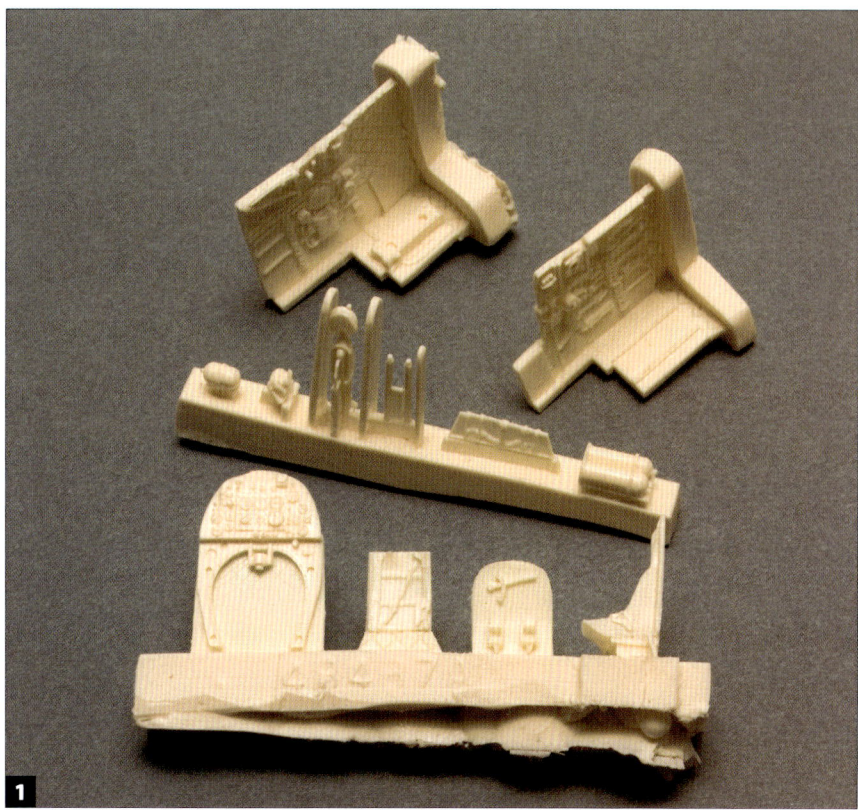

1 The True Details cockpit set (No. 48487) is typical of what resin cockpits have to offer. The detail is crisp and the items are complete. Some (the three tanks located behind the seat in particular) are of questionable value, as they're practically invisible when the model is finished—it seems silly to spend time on items that won't be seen.

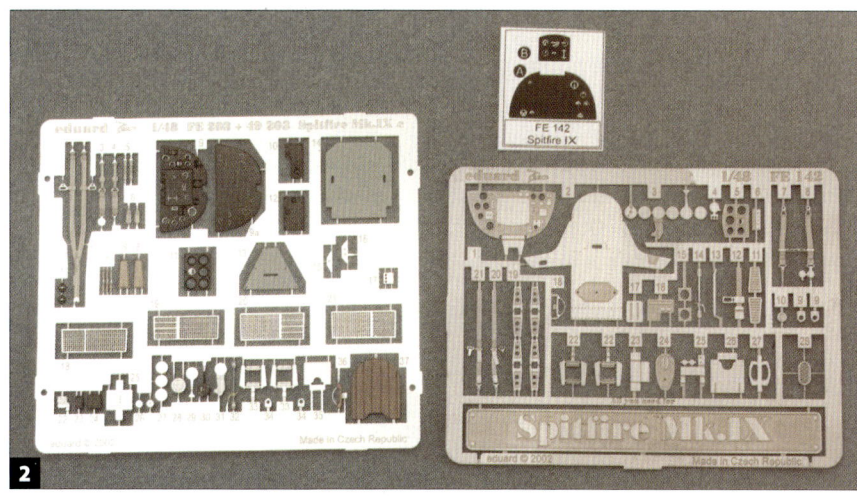

2 Additional detail options are found on a pair of Eduard Spitfire Mk.IX photo-etch sets. On the left is Zoom kit No. 203 with the new pre-painted features and on the right is kit No. 142 in the original bare metal style. Neither is intended for this Occidental kit, but there may be individual items that will work.

Making room. To get an accurate-looking cockpit, a significant modification needs to be made. Between the styrene fuselage and the resin parts I wanted to add, there are essentially two cockpit walls where I only want one, **5**. Here is one of the rare instances where I recommend breaking out a motor tool. With a grinding bit at fairly low speed, slowly wear down the styrene, **6**. Hold the part up to the light occasionally to judge the increasing transparency of the cockpit walls to gauge progress.

After grinding, go in with coarse sandpaper to smooth the ground surfaces and further thin the plastic. With the same coarse paper, start rubbing down the outer surfaces of the resin cockpit walls, thinning them as much as possible without breaking through. By thin, I mean eggshell thin at least. Following this comes test-fitting and adjusting. Happily, once considerable thinning is accomplished, the resin fits the styrene well. A Spitfire didn't really have a floor. Keep the center section to which the rudder pedal bar and control column mount into and discard the rest.

At this point, you can glue the walls in place and paint them and the other cockpit components to a finished state, **7**, **8**, **9**. Conveniently, the fuselage halves can now come together and further cockpit items may be inserted as required. The True Details resin parts set is very complete, but with the Eduard photo-etched sets on hand, it would be interesting to compare some things.

Resin or PE? Both Zoom sets consist mostly of cockpit details. Because all this stuff is present in the resin moldings, it seemed redundant to the project. I did remove the photo-etched rudder pedals from both Eduard sets to compare to the resin interpretations, **10**, but I still preferred the resin both for ease of handling as well as for the size of the parts.

Both Eduard sets also contain multi-piece instrument panels that I put together, again, just to compare to the resin and kit parts, **11**. While the Eduard interpretations provide better dial detail, I thought the painted-up resin piece did a fine job of representing this prominent feature and was easier to fit to the model. The photo-etched seat and shoulder belts looked interesting, but the same items were molded on the resin seat. As with the instrument panel, some careful painting brought them to life nicely and quickly.

There's little reason not to build the Occidental wings stock with only a couple modifications. The wingtip lights are poor-

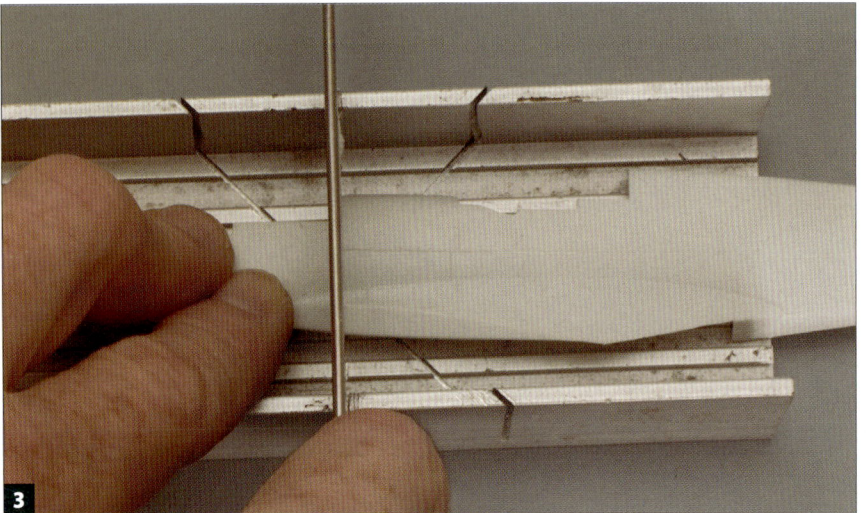

3 The use of a miter box facilitates a straight cut during the removal of the nose of Occidental's Spitfire.

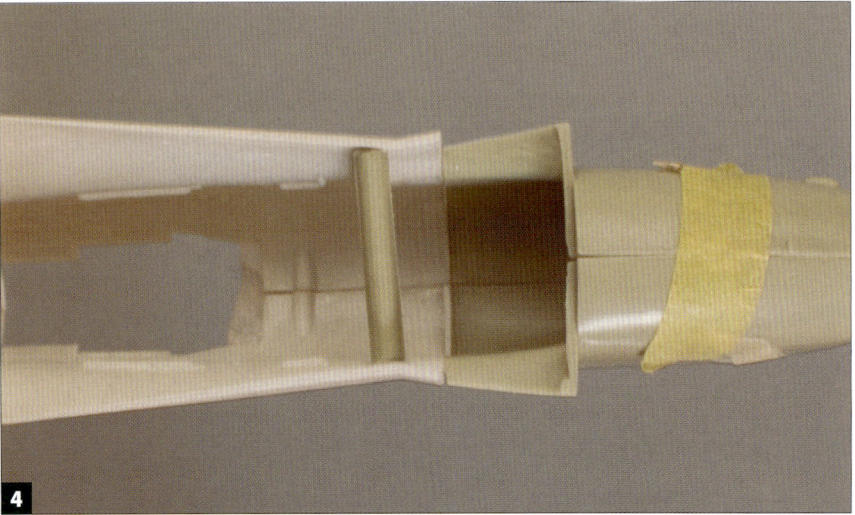

4 With the Occidental nose missing, a spreader bar made from a sprue helps the fuselage align with the walls of the Otaki nose. A significant mismatch of left and right sides is obvious, so adjustments will be in order.

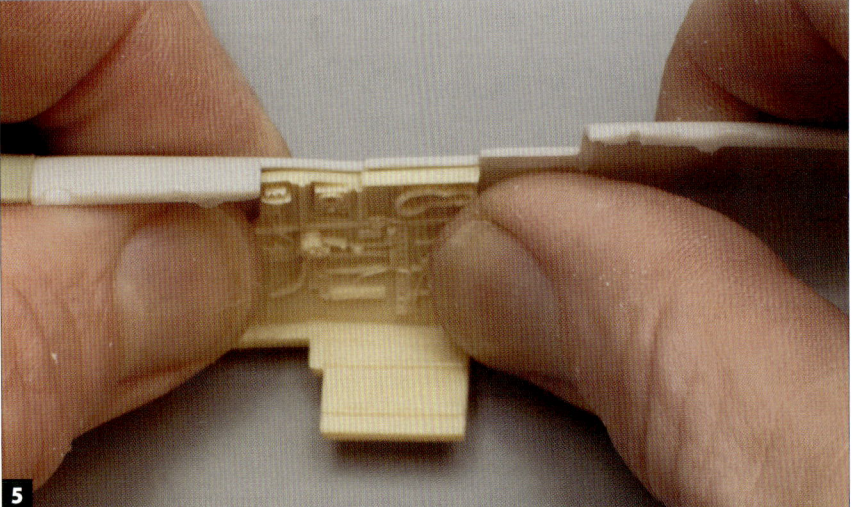

5 Without any modifications, the right side of the resin cockpit fits like so. The wall thickness needs to be reduced considerably.

ly molded, so I substituted tapered pieces of styrene rod, **12**. Also, the shell ejection chutes for the four outboard .303 machine guns were filled. The biggest chunk of work goes into blending the poorly fitting separate wingtip pieces into the rest of the wing. Finally, we need to cut a V from the center wall that will eventually meet the lower nose. This will allow the wing to bend and allows us to crank in some dihedral when it mates up with the fuselage.

At this stage, I want to see what problems the hybrid fuselage-to-wing joint is going to present. Test fitting reveals work ahead, but no major issues, **13**, **14**. As far as kit-bashing operations go, this one in fact looks relatively trouble-free.

The joining of the fuselage halves presents no real fit problems. A round oil-filler cap forward of the windscreen needs attention though. Besides being slightly out of round, it's sunk too low and has an impossible-to-fill seam line through it, **13**. My fix is to drill it out to a slightly bigger diameter and fill the new hole with a piece of slightly stretched styrene tube. In effect, I fill the drilled-out hole with a smaller hole. Then with my punch-and-die set, I punched an appropriately sized new filler cap. I reamed the new hole to the exact size of the tiny styrene disc and press-fitted it in place.

New shoes. Neither the Otaki nor Occidental kit has very impressively detailed landing gear. Fortunately a Spitfire's landing gear is pretty basic. I decided to make my own using a combination of styrene tube for the basic strut, square strip for the mounting peg, and the bottom of the Otaki strut, consisting of the oleo and the axle, **15**.

The Eduard set for the ICM Spitfire kit contains photo-etched oleo scissors. Perfect, thought I. However, they ended up beautifully illustrating a characteristic of metal parts to be aware of. Once removed from their fret, these parts are meant to be folded in half, then bent into their typical V shapes. In the process of bending the second one into a V with a pair of tweezers—thereby making it essentially a spring—it sprung loose.

Tiny lightweight pieces like that with a lot of tension can and will travel a great distance if they escape your grasp. Not only did this piece shoot off my bench, it

6 With a round router bit in a motor tool, the kit wall is ground down as thin as possible. Holding the part up to the light through the process reveals progress.

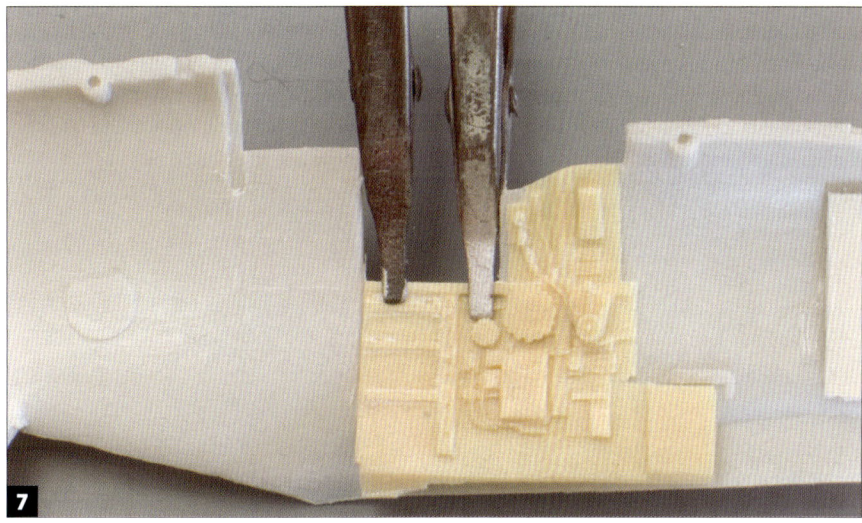

7 The left cockpit wall is clipped into place after the grinding and thinning process. The initial fit looks good.

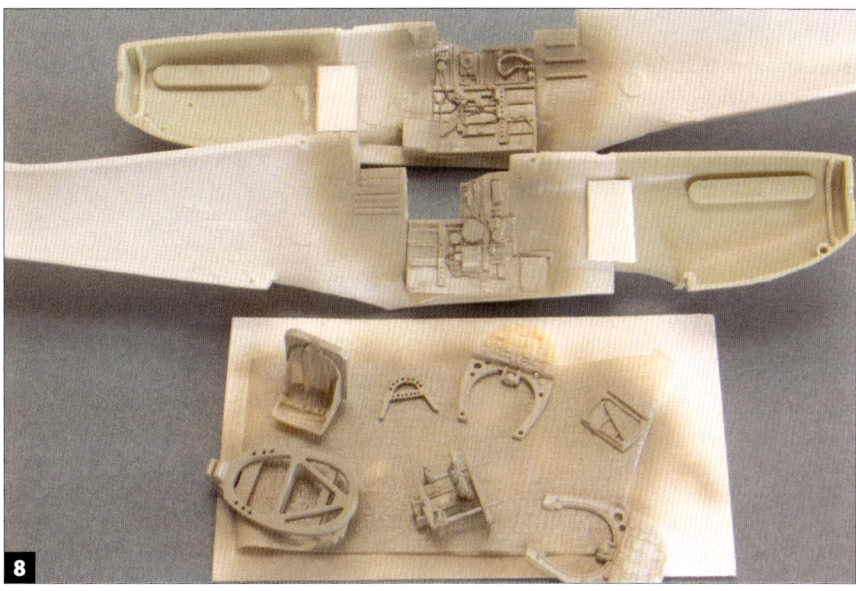

8 The cockpit walls are installed and all interior components are given a base coat of RAF interior gray-green. Substantial reinforcement pieces are seen superglued at the joint of the new Otaki nose halves.

9 Using the methods shown in Chapter 2, a dark gray wash is applied, details picked out, a flat coat applied, and a silver dry-brushing added to lift the raised details a bit. All the components are then about ready to install. One of the instrument panels is the resin replacement and the other is the kit piece for comparison.

10 Here's an interesting rudder pedal comparison. The assembly shows the resin pieces put together. Stuck in the Blue Tack are the Eduard rudder pedals for the ICM kit; the shiny pieces underneath are the rudder pedals for the Hasegawa kit. The ICM pedals look to be 1/32 scale (if not bigger) and the Hasegawa pedals fell apart during the bending process.

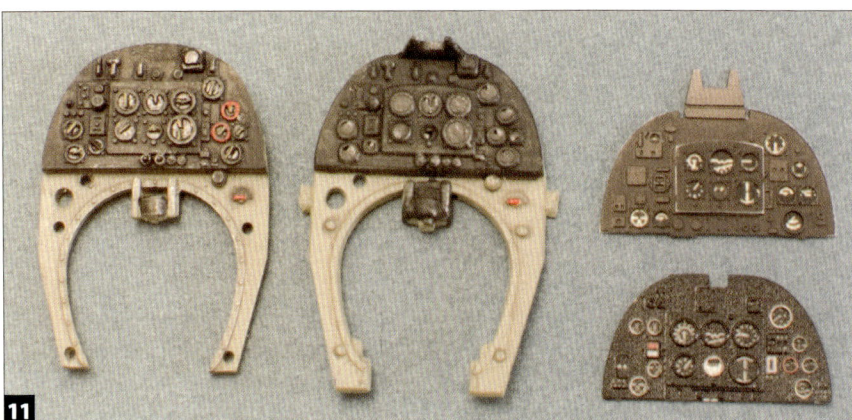

11 From left to right are the painted resin instrument panel, the painted kit instrument panel, the Eduard Zoom rendition for the ICM Mk. IX, and under it, the Zoom rendition for the Hasegawa Mk.IX. The Eduard examples are meant to be glued to kit parts in their respective kits.

took the other bent scissor with it. After an hour's search, I could only find one of them. In this case, one is only good for a template, so I used it as a guide to make a new pair of scissors from styrene.

Both Eduard sets also include tie-down rings for the gear struts. Though a bit oversize to my eyes, I added these on a thin strip of lead foil wrapped around the strut at the appropriate point. Lastly, I glued the struts to the Occidental gear doors and put these finished assemblies in a safe place.

A True Detail kit furnished the wheels to the project, **16**. They represent wheels with the spokes covered, which is a little different look, but one often seen on the late-war Mk.IX's this model represents. Once they are removed from their resin "shoe," smooth out some of the bulged sides and drill the axle hole to accommodate the Otaki axles.

Match-making. The big job of this kitbash is getting a smooth match of the wings to the fuselage.

This will seal the cockpit from underneath, so finishing the cockpit becomes a priority, **17**. Finishing means placement of the control column and rudder pedal assembly removed from the floor provided. The correct height of the control column in relation to the seat and instrument panel is your main concern. To achieve this, attach this sub-assembly directly to the inside of the bottom wing. I ground down the resin on the bottom of this fragile little assembly. It's tricky to hold firmly at the same time you work on it, and I broke off one of the rudder pedals. Get used to it—this is one of the joys of working with tiny resin pieces. Fortunately resin doesn't tend to fly into space the way PE parts can, and I was able to repair the break.

Pinch the solid middle section of this interior piece with a pair of needle-nose pliers to avoid any contact with other exposed pieces, then grind away with a sanding stick. Once you have reduced the piece as necessary, superglue the piece to the wing.

I encountered some interference with the rudder pedals and the rib portion of the instrument panel piece. I broke loose one side of the offending rib, repositioned it, and trimmed away a bit of the other side. This adjust-fit-adjust-fit process well illustrates a typical experience with multi-piece resin detail—rarely does any of it simply click together. Once everything comes together, you can simply brush-paint the bottom of the wing underneath the rudder pedal/control yoke assembly a dark gray-green to suggest depth into the fuselage and to provide contrast with the rest of the lighter-colored cockpit pieces.

The bird takes wing. Now it's time to attach the wing. Glue the rear-most end of the wing (actually the lower fuselage) in place. Masking tape stretched wingtip to wingtip over the fuselage sets the dihedral, **18**. You'll have plenty of wing-root gap to fill with sheet styrene and super glue. Don't be stingy with the super glue, as you want a strong bond that no dihedral tension in the wings could ever crack. Where the front of the wing meets the forward fuselage, you may find both a gap and an overlap with the Otaki part. Overlap is sanded away and the gap is filled with epoxy putty, which is blended and sanded smooth to achieve acceptable contours all around, **19**, **20**. The big oil scoop underneath has to be bent and trimmed to get a good compromise fit.

After all the seams have been filled and smoothed, turn your attention once again to the cockpit area. A small resin gunsight is part of the True Details package. This should be attached as high as possible above the instrument panel so that its view clears the lower frame of the windscreen.

A small rectangle of clear sheet needs to be attached between two little mounting arms. Because the inside framing of the windscreen will be quite visible on this one, mask it off and paint it flat black, **21**. Cutting masking tape to the shape of the outside panels and placing these on the inside surfaces makes this fairly easy. Once painted, the windscreen can be attached. There's a significant gap across the top of the fuselage that will need to be filled with a thin styrene shim.

With this seam smoothed out, the windscreen exterior is ready to be masked off, and you're all set to start painting the airplane.

Finish line. I chose the alternative kit markings supplied by Occidental, which represents NL-L of No. 341 squadron, a

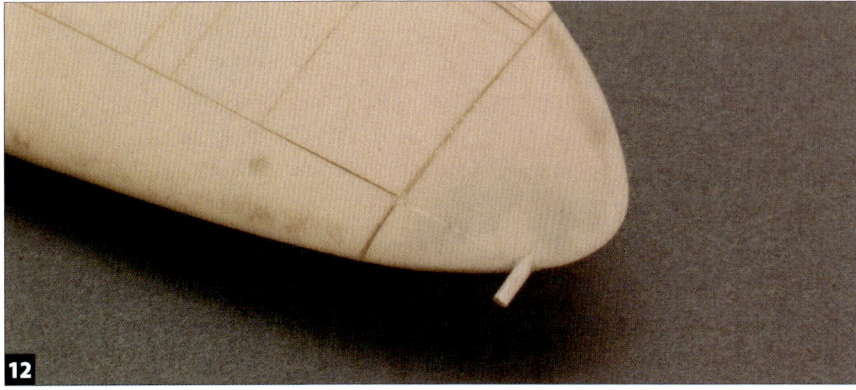

After the wingtip pieces are blended into the wings, the out-of-round navigation lights are trimmed off and replaced with sections of styrene rod. These will be trimmed back substantially before the model is finished.

The wing looks like it will work fairly well with some trimming and shimming. The gap at the left front of the wing is curious, as everything matches well at the fuselage there. The oil filler cap on the top of the fuselage will need some attention.

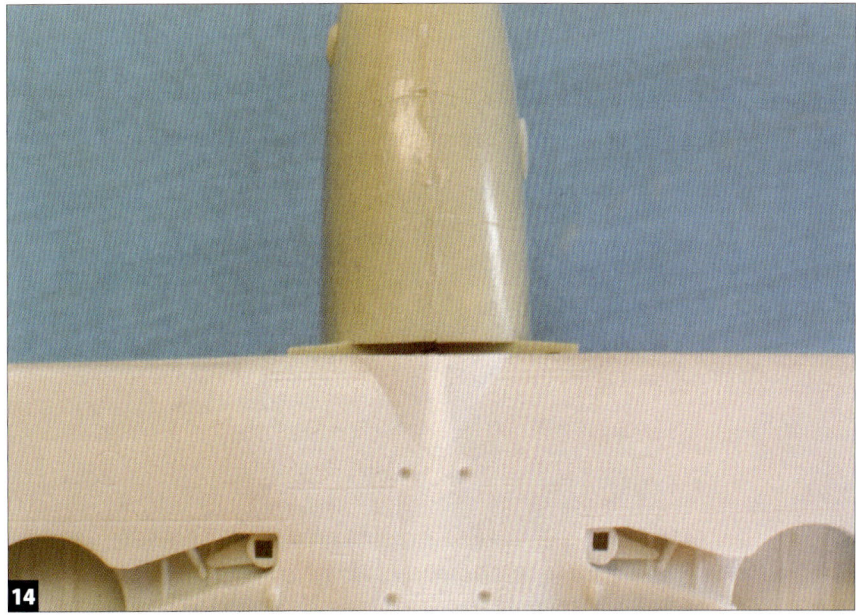

This is easy: Shim the gap, trim the overlap, and fill the resulting holes with epoxy putty.

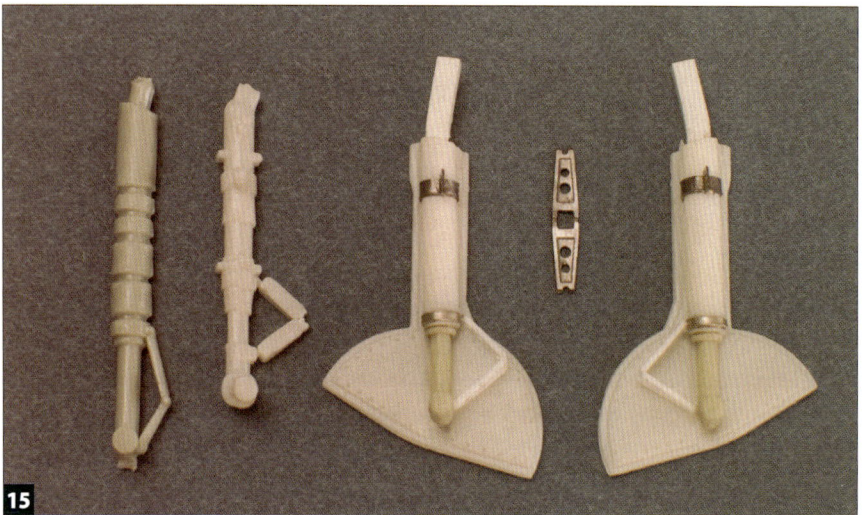

15 Here's a comparison of the Otaki, Occidental, and the scratchbuilt landing gear. The single PE oleo scissors was used as a template to make a pair from sheet styrene.

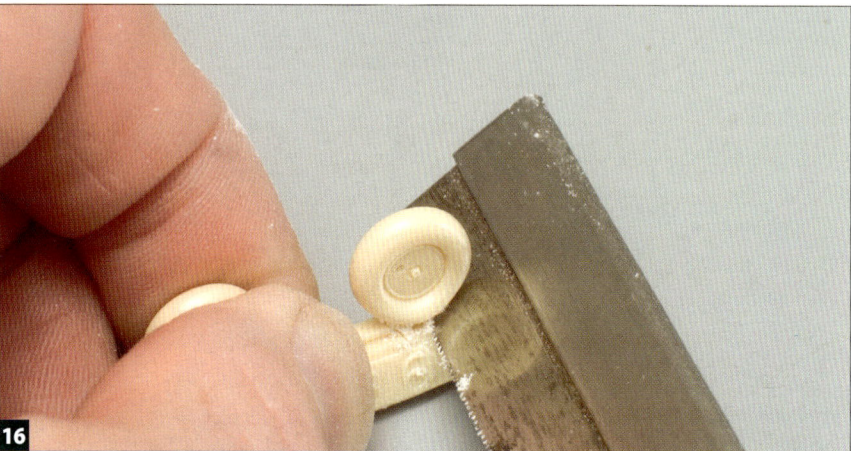

16 The True Details resin wheels are removed from their shoe with a V cut underneath them with the razor saw. Once removed, they are cleaned with a sanding stick. The sides are also sanded down some to reduce the squashed-bottom look, typical of these resin "weighted" tires.

17 At this point the wing is ready to be attached. Just forward of the instrument panel, a styrene bulkhead has been installed to represent a blanking front wall of the cockpit. It's painted the same darker gray-green as the bottom of the wing.

Free French unit with the RAF. There's a photo of this machine on the kit instruction sheet, but all it really shows is a well-used Spitfire. Further research didn't turn up significant information. French Spitfires in the RAF don't seem to have been well documented.

After painting the model in the standard British late-war scheme of light gray, medium gray, and dark green, I put on the kit decals. The wing roundels looked funny to me: The separate blue center dot looks too small. However, with no photos of what a French roundel on a Spitfire wing actually looked like, I decided to go with my instincts for how it should look. I cut bigger dots with the help of a scribing template for circles. Then it occurred to me to question when these roundels would have been applied and what would have happened to the bigger British roundel that was originally there. Assuming it had been painted out, I cut a couple more circular masks and covered the decal roundel. I then painted around it with a different shade of gray and green in the diameter of a British marking, **22**, **23**. Is it accurate? Maybe. It's certainly a possibility, and it adds a little visual interest.

After the Allies began establishing bases on the European continent, the British began to paint out the Sky spinners and fuselage bands on their fighters to make them less conspicuous on the ground. The back spinner on this Spitfire dates it to this period. The supplied photo shows the propeller, but not the fuselage, so it may be safe to assume the rear fuselage band was overpainted too. Again for the sake of visual interest, I painted a band in a slightly different shade of green, suggesting the ground crew found some suitable paint to cover the original light band back there.

Absent from the Occidental decal sheet is a serial number for this machine. What NL-L's serial number may have been is anyone's guess. I found a list of Spitfire Mk.IX serials and learned that from regular production numbers and conversions from Mk.Vs that there are almost 6,000 to choose from. I applied only the parts that were conceivably not on the Sky band and therefore not overpainted as well. Judging by many different Spitfire photos, those serial numbers didn't get a lot of respect. Wing walk stripes and other minor stencil-

18 Masking tape pulls the wingtips up where they belong, and sheet stock is fitted into the biggest wingroot gaps.

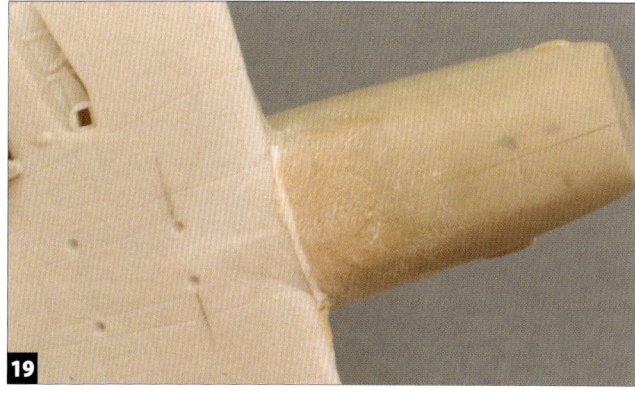

19 More epoxy putty is employed to blend a step in the Occidental portion into the Otaki item. The good news is that a big oil cooler scoop will be positioned right here, making a perfectly smooth transition unnecessary.

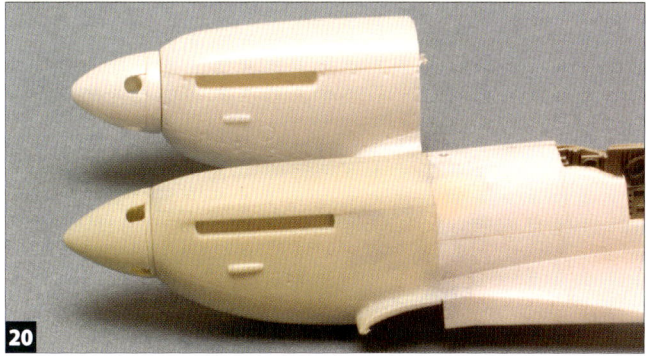

20 Is the nose swap worth the trouble? I suppose it depends how well you know the look of a Spitfire. I don't count rivets, but I do like to see the proper shapes. The slight tan area on the fuselage is epoxy putty making a smooth transition at the most serious disagreement of the Occidental and Otaki contours.

21 A thin shim is required under the front of the windscreen, and the new oil filler cap can be clearly seen. The wingroot gap is gone, filled with a combination of styrene shim, super glue, and auto primer.

22 **23** Circular masks cut from self-adhering notes are used to suggest where the British roundels were located before being replaced with French cockades. A slightly different shade of gray and green subtly changes this round area.

ing come from an Aeromaster sheet, as the Occidental items look a little oversized.

After flat coating, dark gray and black pastel chalks were rubbed on around the cowling area, the wingroots near the cowl, and around the big blistered gun panels on the wings. Underneath, black and brown artist's oil paints were streaked down the belly from behind the engine firewall area. A compass with a blade installed helped cut thin circular strips of Tamiya masking tape, which was wrapped around the fronts of the bombs. With a slightly raised guide in place, I carefully painted the tape yellow.

The last details were the lights. I painted the one on the rudder white, then carved, sanded, and polished a piece of clear stock to a half-teardrop shape for the light behind the antenna mast. On the wingtips, I touched the fronts of my replacement light fairings with a tiny dab of clear epoxy and let that cure. I then carefully painted the clear bumps dark red on the left and dark blue on the right.

There it is—not a perfect Spitfire, but a pretty respectable one, I think.

HANDLING VACUUM-FORMED CANOPIES
1/72 scale P-40N Warhawk

Vacuum-formed canopies better resemble the thin cross-section of the real thing, which is especially important when modeling them in the open position.

Contrary to what you might believe by looking at most canopies, observation domes, and turrets in plastic kits, these items are not thick chunks of armored glass. In reality, they were generally quite thin in cross-section. There's really only one way to achieve this scale thinness in your models (especially in smaller scales), and that's with vacuum-formed clear parts.

An excellent example of such "vac canopies" are those in the Squadron/Falcon line of Crystal Canopies for individual subjects and Falcon Clear-Vax sets for related groups of airplanes. These sparkling little jewels look impressive on their own, but blending them onto a model airplane usually calls for precision work.

Made to fit—with a little help. The subject of this exercise is Hasegawa's 1/72 scale P-40N, originally released in the early '70s. It comes with a single-piece closed canopy that fits reasonably well. It's fine unless you want to dress up the cockpit a bit and show it off. If this is the case, by far the easiest solution is to replace the kit part. Enter Squadron item SQ9133: two beautifully formed P-40N canopies. The suggested kit for this pair of clear parts is the Hasegawa offering, which leads one to believe it's been made to easily fit this particular model. Let's see, shall we?

Step one is to remove the canopy from its backing sheet. With a sharp pair of scissors, trim away as much of the excess as you dare, **1**. Next, with a fresh single-edge razor blade, trim further, peeling away more backing material. Make sure that pressure on the blade is always going away from the piece and not toward it, as a slip can turn a canopy to scrap in an instant. Stick the canopy on a cutting pad with a looped piece of masking tape. With the canopy secured and using the razor blade like a guillotine, chop away as many straight sections of excess stock as possible, **2**. Like every step from here on, a slip or a crooked cut can't be easily fixed.

After the straight segments have been taken care of, perform the same chopping process to whatever extent you can to the curved rear section, **3**. The balance of the trimming can be accomplished with a medium-grit sanding stick. (It's a good idea to protect the canopy with masking tape if sanding sticks are in play.)

Two in one. Sometimes these vac canopies are formed in separate sections, sometimes not. In the case of this Warhawk, you get both. The windscreen and sliding hood are one piece, and the rear section is formed separately. There are two approaches to separating the windscreen from the sliding section. When a pair of canopies is provided, my guess is that you are to sacrifice one for the sliding section and the other for the windscreen. However, if your hand is steady and eyesight good, it's possible to chop fused sections cleanly apart with your razor blade. If you make a mistake, though, it will negatively affect two pieces.

Now that you have the canopy stuck on its side with masking tape, reach with a razor blade as far into the canopy as the

1 A pair of sharp scissors cuts away most of the backing plastic. You can't beat the control of a good pair of scissors.

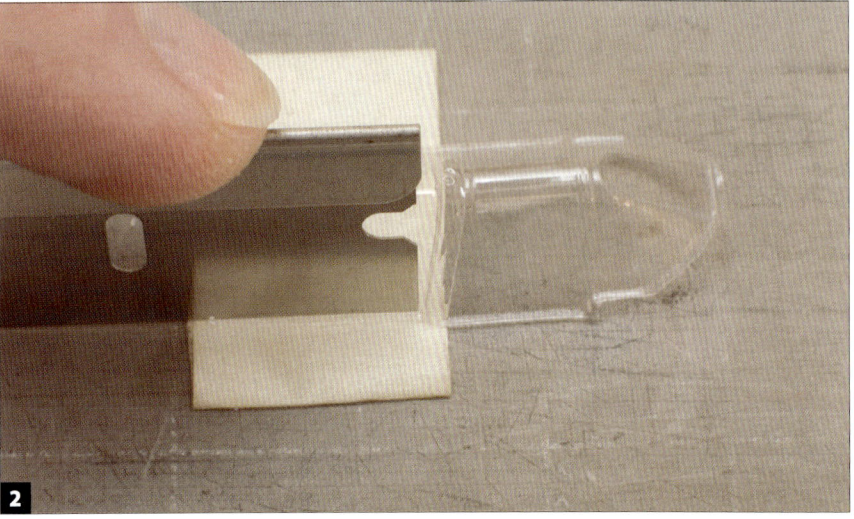

2 With the canopy stuck to masking tape, a razor blade is applied like a guillotine flush with the canopy to chop away straight sections of the backing plastic.

3 To get a straight clean separation of the windscreen and sliding hood portion, the edge of the razor blade is inserted as far into the inside of the canopy as possible, lined up exactly on the dividing line between screen and hood, and another guillotine press is applied.

flat side will allow. Line it up exactly with the slight space between the framing of the windscreen and hood and apply pressure on the blade with some emphasis on the forward edge so the point cuts first. If nothing moves other than the blade, a quick, straight cut results.

Repeat this on the opposite side of the sliding hood section, which will leave it connected to the windscreen only at the top curved part. After sticking the piece upside down on the masking tape, slide the razor blade into the spaces opened between canopy sections and gently press straight down. The two pieces should be free of each other, **4**. Again, fine-tune with a sanding stick as necessary.

Test-fitting the canopy to the fuselage before the fuselage is glued together reveals whether this project will work, **5**. In this case, the fit is acceptable, so the canopy pieces can be dipped in Future and set aside while the building of the model proceeds. If the canopy is too big, **6**, a minor widening of the fuselage with shims would be an option. However, this will surely cause problems elsewhere that may not be worth the trouble.

Sticky decision. About the time the painting stage is reached, it's time to start thinking about the canopy again. Set it in place and make any last fit adjustments. The gluing options are these: If the fit is perfect, white glue might work. The advantage is that it won't damage the clear part if any finds its way onto the canopy. The disadvantage is that a perfect fit is hardly ever found, and as a result you will have gaps to deal with.

Gaps can be filled with more white glue, but white glue won't allow the smooth blended-into-the-fuselage look that is important to me. Besides that, the framing still needs to be painted. That means masking tape on the canopy that will need to be peeled off. There is precious little gluing surface between the razor-thin clear parts and the rest of the airplane. Will the white glue hold the canopy in place under that pulling strain? Probably not.

The other option is super glue. The advantages are that the canopy will be permanently attached, which will allow filling and blending into the airplane skin where necessary, so lifting of the even-

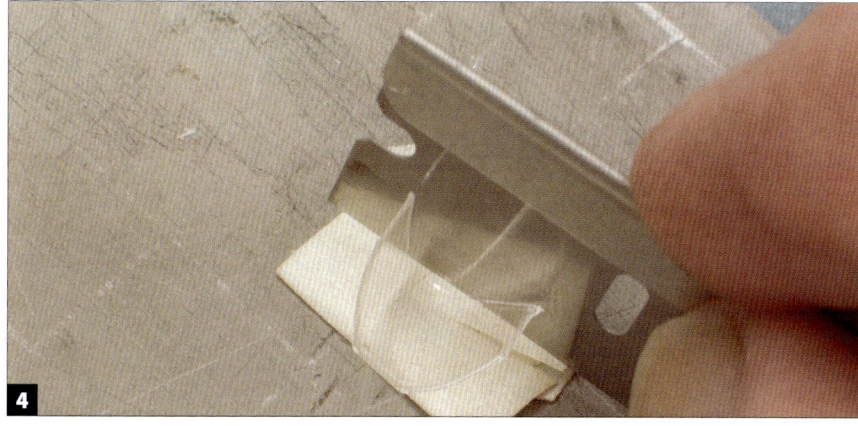

4
Having done this to both sides, the canopy piece is stuck upside down on the masking tape, the razor blade is inserted into the fresh separations and pressed straight down again, making a third and final cut.

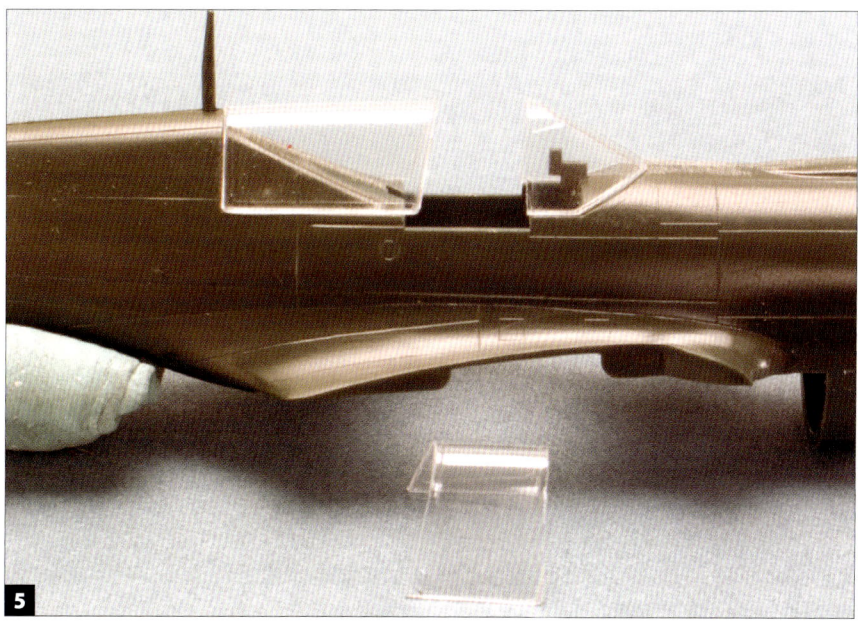

5
With the three pieces separated and trimmed, test-fit the piece to the fuselage. Things look good, except for the rear piece being a little too long.

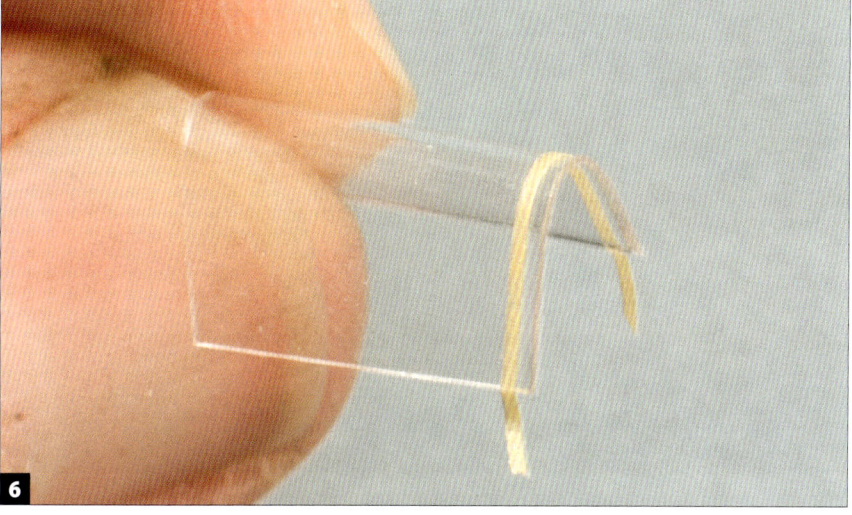

6
To accurately trim off the excess length, a thin strip of masking tape is used like a new piece of canopy frame. This will provide the accurate guide needed to trim the piece.

7 Attaching the windscreen begins with securing the corner edges in place. This reveals a gap on the right side.

8 A small dab of epoxy putty is smeared and blended into the gap, giving the look of a cream-colored seal.

9 With the clear parts masked to guard against scratches, the primer-coated seams are carefully wet-sanded with the help of a balsa sanding block. The edge of the sanding block forces the sandpaper into all the corners.

tual masking tape shouldn't cause any concern. The disadvantages are potential frosting of the clear parts (the Future dip should eliminate that) and the possibility of getting glue on the canopy. Although white glue will peel off when dry, expect super glue to be on there for good.

Fine work with super glue. The best fit of the windscreen still leaves a bit of a gap at the right front—so much for the white glue option. After carefully super-gluing the corners in place to keep the piece from shifting, **7**, continue to sparingly apply fast super glue with the end of a piece of stretched sprue all around the surfaces that touch the airplane. Get just enough of a bead on the end of your sprue "wand" to transfer from it to the seam between canopy and airframe. The capillary effect of the super glue takes it along the edge, and only along the edge.

On this P-40, there is nothing to interfere between the edge of the canopy and the surface it needs to attach to. If something else were touching low inside the windscreen, you can assume the super glue would travel from the point where it's deposited to the seam it's intended for and from there, directly to whatever may be touching the canopy from behind. Prepare to gasp in horror as the super glue proceeds to run up the inside of the canopy finding yet another gluing surface to connect. Such an event is a guaranteed to spoil your day.

The gap in front isn't big enough to shim with styrene, but it's too big to feel comfortable about the amount of potentially frosting super glue needed to close it. Here, I would choose to carefully stuff it with epoxy putty, **8**. When the putty cures, mask off all the clear areas for protection. With all the precision you can muster, gently sand the canopy-to-airframe seams. After a coat of primer on the sanded areas, wet-sand once more to smooth things as much as possible, **9**. Remove the masking tape for final inspection, **10**.

Paint the frame. The final hurdle is painting the framework. In the case of this Warhawk, there isn't a lot of that to do. That's the good news. In 1/72 scale, what we have to deal with are some very tight, thin curves, which are especially tricky on the windscreen.

Mask off the main curved frame on the front of the windscreen. The Hasegawa kit windscreen piece is a good guide for the shapes of masking tape needed there. Tight oval shapes on scribing templates are a great help to get the tight curve needed on top. The sliding hood portion, being all straight lines, is easy to mask. The rear fixed section has a straight frame on its leading edge and two very thin frames in the middle. In order to get these as straight and even as possible, cut thin strips of masking tape the width of the frame lines and lay them on top of the clear part. Eyeball and adjust this tape to perfect straightness, then use it and not the lightly molded frame marks on the canopy as your masking guide. Peeling off the thin strips of tape reveals the thin straight frame area to be painted, **11**.

An Aussie "Cleopatra." I chose to model the Australian scheme provided in an old Minicraft issue of this kit, so I mixed a green from Extracolor enamels to arrive at my idea of what Australian Foliage Green would've looked like on an Aussie Kittyhawk in 1/72 scale. My shade of green ends up being a good match to FS 34096. The gray is Extracolor X158 Air Mobility Command Grey. Photos of "Cleopatra" suggest there is little difference in tone between the gray and the green.

Photos also suggest there was a painted-out roundel or something behind the obvious one. With a blade mounted in a compass, I cut two roundel-sized white decal film discs and painted them a slightly lighter shade of green than the rest of the airplane. I also scribbled this mix randomly over the initial base coat to give it an uneven look. The ailerons were fabric covered, and it's likely they weathered in the harsh sun differently than the rest of the aluminum skin, so they received this lighter mix as well.

The decal part of this project illustrates a couple of important factors. On the sheet, the white markings look to be more of a cream color than true white. Unfortunately, the white ink was not printed with much density, so when these cream-colored decals hit the dark green, they become light green. This contrasts sharply with the white leading edges of the wing and tail surfaces already part of the paint job. Fortunately, I had a second kit and

The masking tape has been removed to check progress. All looks well, so the canopy can be re-masked for painting.

To exactly place the two thin frame lines on the rear section, they are first covered with thin strips of tape. With those perfectly placed, I mask around them. That complete, the thin strips are removed, exposing the canopy to the paint that will make up the frames.

A second BU squadron code decal is applied on top of the first one to compensate for the lack of density of the white ink of the decal. The individual aircraft letter has already been "doubled up."

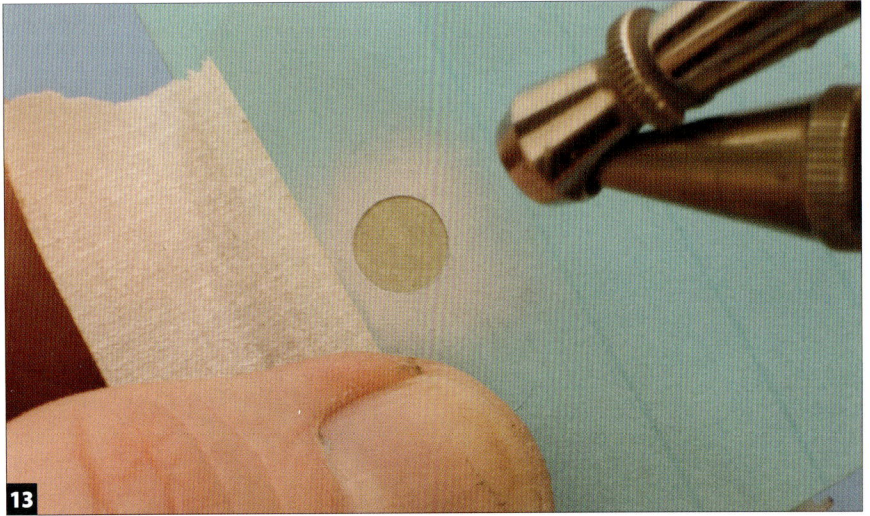

13 To get a truer white on the national insignias, a disk was cut out of a self-adhering note to match the white of the roundel. This was pressed on the decal, and a coat of white paint was airbrushed onto it. Note the decal's green appearance compared to the white overspray on the mask.

14 The Warhawk received a minimum of weathering, with just a bit of wear on the leading edges of the wings and atop the wings in areas of foot traffic.

decal sheet on-hand. (Never underestimate the value of having a back-up kit!) Layering a second set of decals over the originals lessens the problem, **12**, but doesn't really eliminate it.

My options at this point are to live with the incongruity, mute down the already painted white to match the decals, or add a third layer of white to the decals. I chose the last. I cut a mask from a self-sticking note with the help of a round scribing template to match the white of the roundels, **13**. Conveniently, they're all the same size. I stuck this on the six positions and gave them a good shot of gloss white.

Cutting masks for the code letters would be a pain, to say the least. Instead I used my best pointed paintbrush and some of the same thinned white paint to dab and drag a white wash over the letters. It's precise brushwork, and slips have to be either scraped away with a No. 11 blade or re-touched with the Foliage Green mix. They may be a little uneven on close inspection, but they match the rest of the white markings, which was my higher concern. I left the 29 on the chin and the "Cleopatra" alone. One needs to know where to draw the line.

Flat coat and weather. The model is flat-coated and weathered with restraint, **14**. A common habit among modelers is to apply long black streaks leading back from wing-mounted machine guns. On a Kittyhawk having white leading edges, this sort of thing would be very prominent, yet after going through all my reference photos of Australian machines (and some similarly marked U.S. 49th FG machines), I could find no evidence of such long stains. Shadows and battered wing leading edges are in evidence, but no long, sooty streaks. So either I didn't find the right pictures, ground crews were quick to clean up their airplanes, P-40s never fired their guns, or when they did fire, they didn't make quite the mess often suggested. Whatever was going on, I believed what I saw in my photos and didn't add significant streaks to my wings.

Typical chipped paint on a 1/72 scale model should be hard to see except in areas of high traffic. Where would shoes and boots actually come in contact with the airplane? Probably only where the pilot would get in and out and where the ground crew would service the machine. The most-serviced areas would probably be the engine, guns, and fuel, oil, and coolant access areas. Panels around these areas are bound to get dinged and scratched in opening/closing and removing.

How big would these dings and scratches be in 1/72nd scale? Leading edges of the wing would probably take a beating via prop wash and exposure to fuel hoses dragged over them and the like. Unless there was some kind of battle damage, most of the rest of the machine would be largely left alone. If there was some repair to the airframe itself, this would probably show up as a different, fresher shade of paint.

A worn airplane looks the way it does for a number of reasons. I like to study photos and think about the reasons behind what I see before trying to replicate that in miniature. I want to represent a shrunken real airplane, and the finish is key to that.

Final details consist of placement of a pitot tube from stretched styrene tube, antenna mast from sheet stock, landing light from a tiny disc of chrome tape punched out with the punch-and-die set, then covered with a dab of five-minute epoxy. The molded-on navigation lights got just a touch with gloss red on the left and gloss dark blue on the right and gloss light gray on the tail.

Cleopatra III
29
B
A29-629

BU

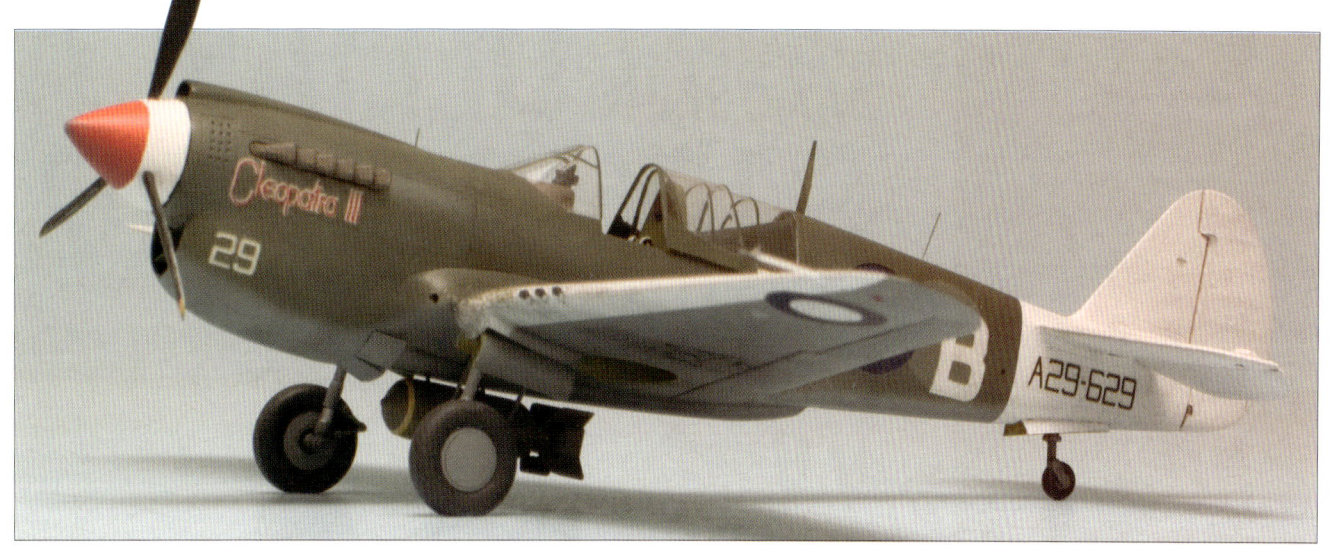

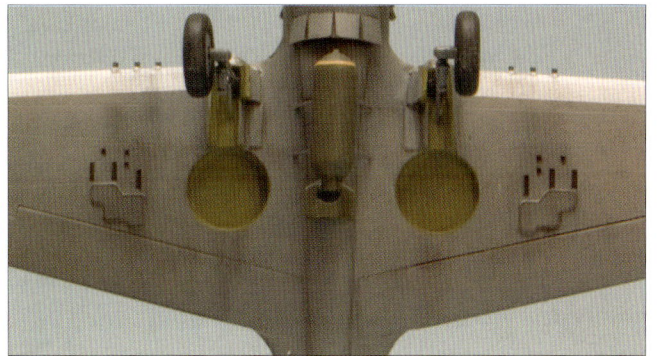

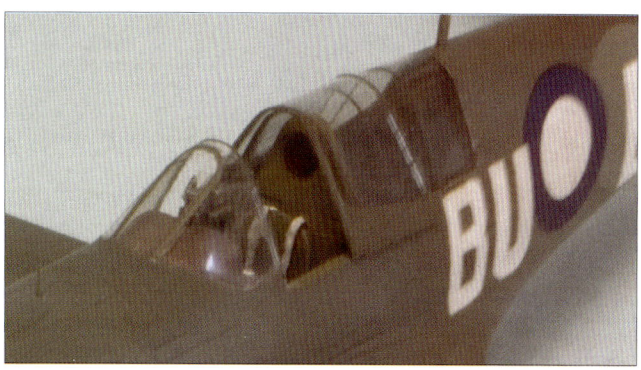

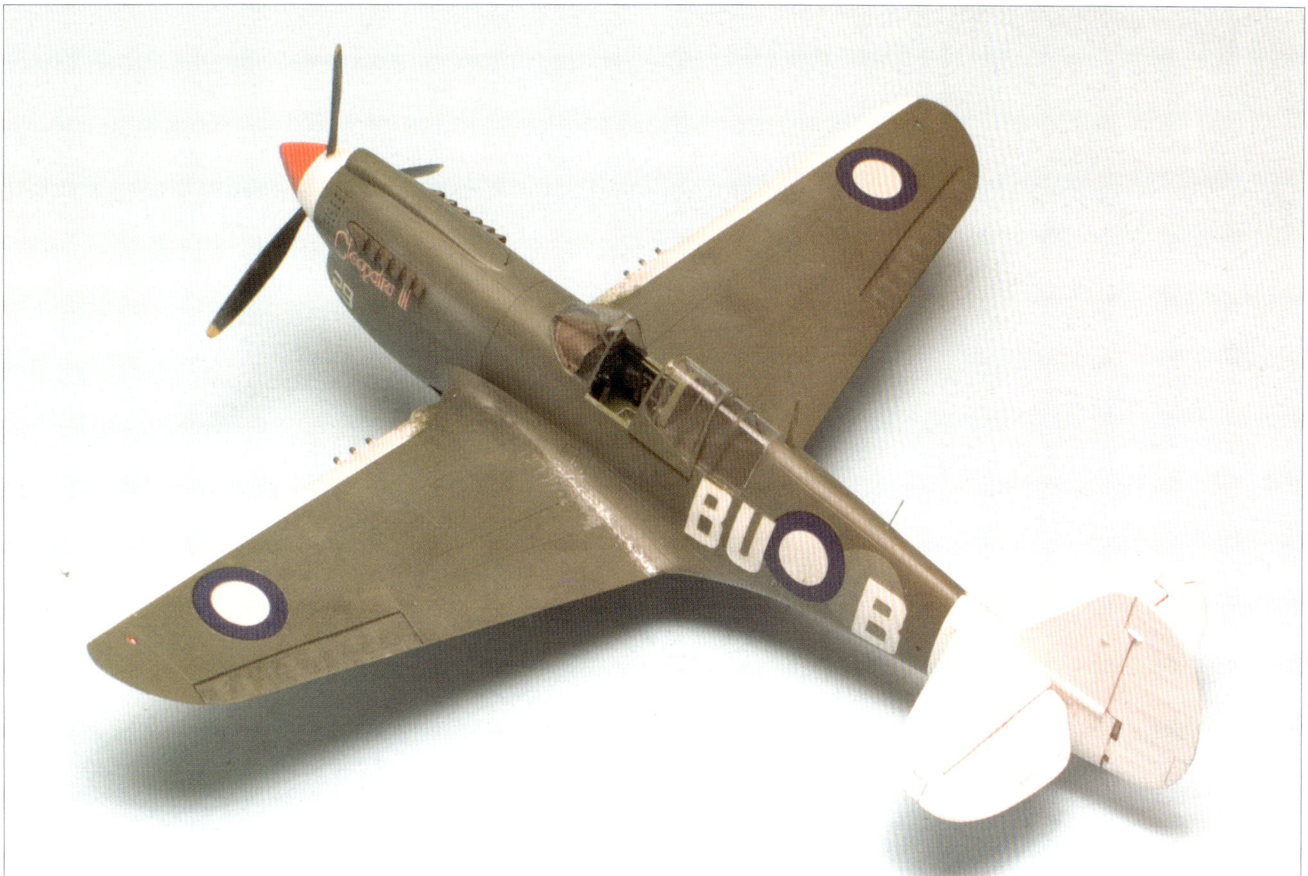
Cleopatra III
29
BU
B

"Vac" and resin conversions

1/72 scale DELTAS F-106B & TF-102

When a kit manufacturer considers spending a lot of money to create molds for a plastic model kit, the ability of the subject to stretch into many versions and "boxings" is heavily weighed.

Back in the days of raised detail, it was nearly a certainty that the manufacturer would only visit the chosen subject once. If there were multiple versions to choose from, the lucky one would probably be the version that had been produced in the greatest numbers by Grumman or Avro or Mitsubishi. This would offer the widest variety of markings options.

Of course when there is more than one version of a subject to choose from, those that don't make it into kit form become all the more intriguing. Fortunately, many enterprising aftermarket firms have produced conversion kits to fill these gaps. Some are as simple as a solid replacement nosepiece to reflect a different piston engine or a camera configuration rather than guns. Others are complete kits in themselves.

Vacuum-formed for variety. The first serious conversion kits to appear were vacuum-formed models, which are relatively inexpensive to produce. Unfortunately for the manufacturers, there has always been a kind of "vacuphobia" in the hobby. One look at that white sheet of plastic with its airplane-shaped lumps, and modelers' minds seem to shut down. Maybe it's the fear of the unknown that freezes their otherwise perfectly adequate model-building skills.

The good news is that resin has stepped in as a popular medium for conversion kits. With the amount of detail achievable with resin and vacuum-formed plastic, you can almost certainly convert your type X to a type Y faster and easier than ever—in theory, anyway. Each type of conversion has its strengths and weaknesses.

Vacuum-formed pieces are heated and stretched over solid masters. The highest point of that master is where the plastic will be stretched to its thinnest.

With a bit of care and patience, realistic models can be built from vacuum-formed kits and conversion components (the F-106B, left) or resin parts (the TF-102 at right).

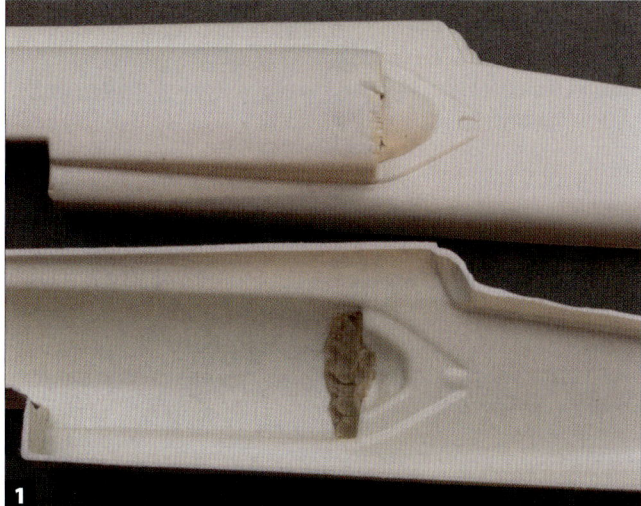

The thinness of the plastic, poor packaging, and rough handling resulted in damage to the edge of the intake bulge. Epoxy putty pressed into the area from behind will press the area back out to its approximate shape. When cured, the hardness of the putty will allow any further shaping or filling on the outside surface.

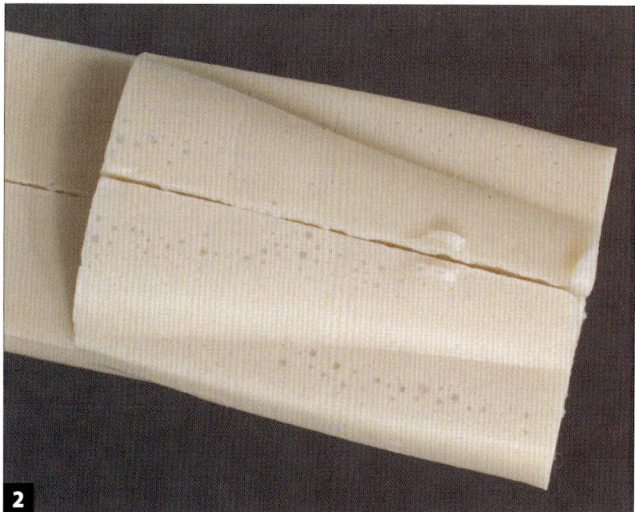

Lurking just under the surface is a bad case of bubbles. As long as they don't touch the surface, there should be no trouble. If they do touch the surface, they become nasty little pinholes.

In the case of pointed parts like spinners or sharp ridges around an opening like a wheel well, the plastic may be stretched very thin indeed. If the kit is not carefully packaged, these high points can be easily damaged by the time the kit gets into your hands, **1**. Usually this kind of damage can be repaired with epoxy putty pushed into the dinged area from behind. When the "fill" is cured, the exterior can be sanded, puttied and sanded again to the correct contours.

Another downside to vacuum-formed models is the impossibility of making any fine detail pieces with this method. Usually the manufacturer tries to provide every part needed, but many of the small parts will only be useful as templates for replacements that you scrounge up or make yourself. The most recent generations of vac kits come with white-metal detail parts such as landing gear, engines, and propellers. These parts are generally better than nothing, but still are best

83

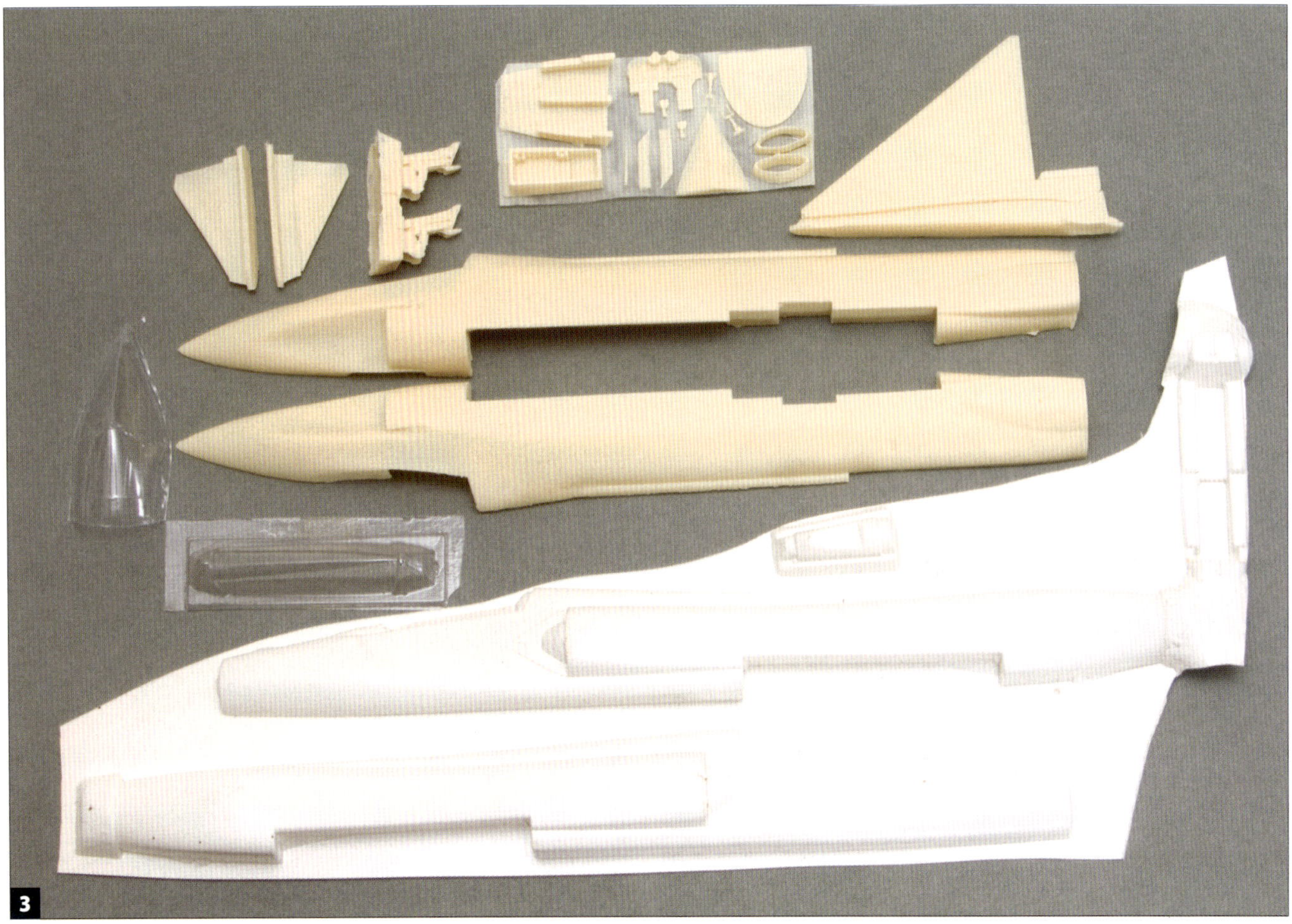

Even though the resin TF-102 conversion (top) is more easily recognizable as a model kit, the simplistic vacuum-formed F-106B underneath is just as viable in its own right.

thought of as templates. I think it's safe to say that vacuum-formed kits require more effort and ingenuity than resin.

Resin rescue. Unlike vacuum-formed models, resin kits and conversions are nearly ready to go when you open the packaging. There will be far less for the modeler to come up with—unless he or she chooses to. And if you like solid material to work with, resin is the answer.

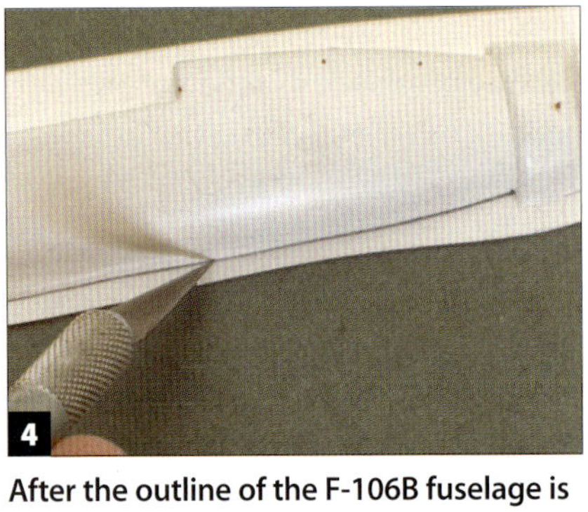

Resin's main weakness is the bubbles that are naturally part of the resin-pouring process, **2**. Though not always visible or significant, if air bubbles touch the surface, or if the surface is sanded down to their level, you get those dreaded pinholes. You can try smearing putty over them, or paint them with primer to fill them, but you'll have to sand this smooth again, risking more pinholes. The only really effective way to fill a pinhole is to drill the bubble out like a dentist repairing a cavity and fill the resultant hole. A bad case of pinholes can create a lot of extra work. If the bubbles end up in thin areas like trailing edges of wings, you'll wind up with a Swiss-cheese effect. The only practical fix here is to replace the bad section with styrene.

Resin's other drawback is its tendency to warp, especially when called on to capture a long, thin shape like a wing or a propeller blade. Heating the part in boiling water until it's softened and bending it to the proper shape may work. I have encountered resin that will bend easily enough, but it always returned to the original warped condition. If you're lucky, that expensive resin kit you buy will be bubble- and warp-free.

Remember that plastic cement has no effect on resin. Your gluing options are super glue or epoxy. I'd strongly suggest super glue as your primary glue tool. Epoxy will let you adjust the fit of parts for a longer time, but that's its only advantage.

After the outline of the F-106B fuselage is traced with a fine marking pen, a No. 11 blade traces that line a couple of times.

Once the line is scored, bending the backing sheet breaks the part free.

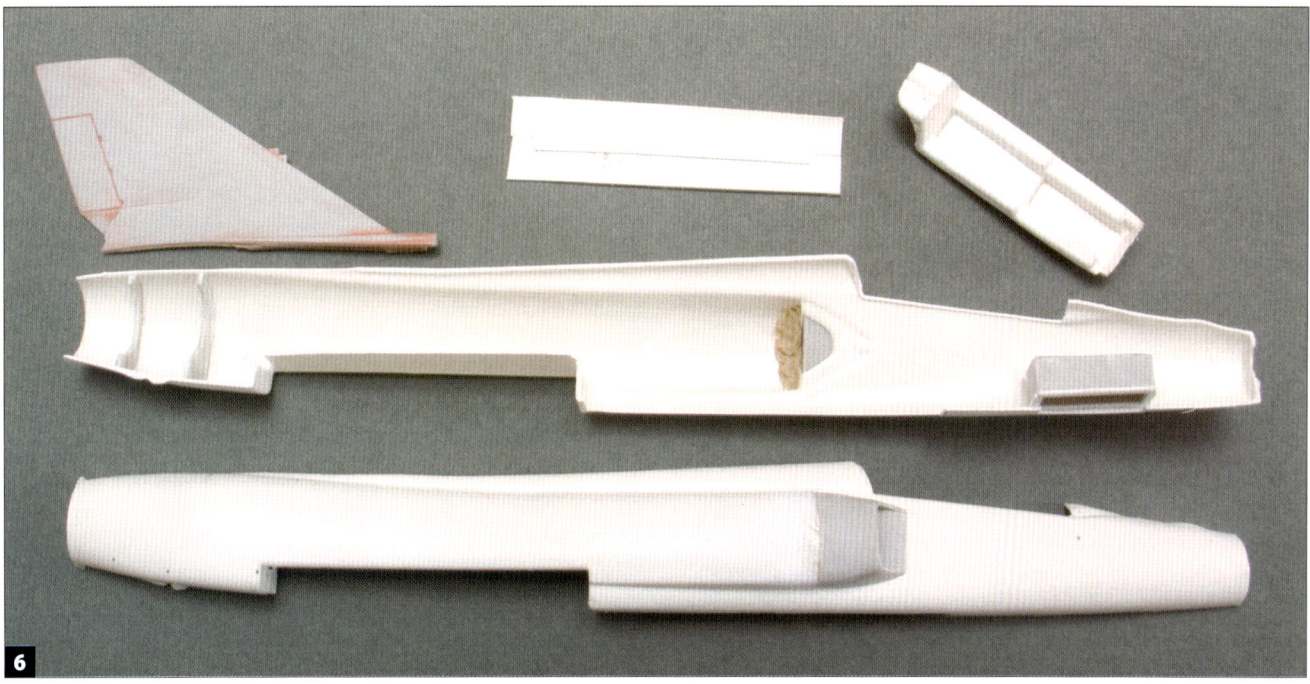

6 Seen here, the Hasegawa intake piece and nose wheel well are attached, and the ribs to hold the engine exhaust piece in place are also removed from the donor kit and incorporated into the vacuum-formed model. The vertical fin is entirely from the Hasegawa kit and will be attached after the fuselage is securely together. Also shown is the section of removed belly (roughly the missile doors), joined and reinforced with strip styrene in the center, and the cockpit tub, such as it is.

Demonstration team. Let me demonstrate both resin and vacuum-formed media with a pair of nearly identical subjects, **3**. One of the most common types of conversions is changing a single-seat airplane into its two-seat counterpart. The Delta Dagger and Delta Dart are perfect examples of fighters that were built in some numbers with a passenger seat. Hasegawa produced the single-seaters in 1/72 scale a year apart in the late '60s. Both kits are good representations but are very basic. The modeling approach and parts breakdown are almost identical.

A few two-seat conversions have been produced for the odd-looking TF-102 over the years, but by far the best is a recent resin kit by RVHP of the Czech Republic. Not only did the company get the shapes almost perfect, it provided much better interior parts than Hasegawa could have even dreamed of in 1969. Whereas the TF-102 had a totally redesigned front end, the F-106B simply replaced some internal fuel tank with a second cockpit and lengthened the canopy. Falcon Industries in Australia produced a perfectly adequate representation of the "Six Bus" in the late 1980s as part of its Triple Conversion III kit.

Both the RVHP and Falcon conversions provide complete new fuselages and the all-important vacuum-formed canopies. The resin conversion features exquisite recessed-line detail, which is in sharp contrast to the raised lines and rivets given in the Hasegawa F-102 kit. The vacuum-formed F-106B fuselage on the other hand is devoid of any surface detail, while the Hasegawa 106 features raised detail identical to the 102's. This means that for both, the conversion and styrene donor kit will need some kind of adjustment to get all the exterior surfaces to match. My solution to the surface discrepancies will be to sand away or fill all the surface detail on both models and start over from there. Rescribing is certainly an option, but these airplanes in 1/72 scale would reveal only faint panel lines at best. Instead, we'll suggest such surface detail with paint at the end of the projects.

F-106B

Break with tradition. The traditional way of separating a vacuum-formed part from its backing sheet is to trim around the part in question, leaving about 1/8" of the backing sheet surrounding the piece. This piece is then rubbed on a flat mounted full sheet of sandpaper until the backing surround gradually thins to nothing and flakes off. This works, but you must pay constant attention to your pressure on the piece in order to sand all surfaces evenly. This is especially true if the piece is particularly long.

There's a much faster method. With a fine-tip permanent marker, trace around the edges of part to be dealt with. Then with a No. 11 blade, trace and score the angle between the part and backing sheet a couple of times, **4**. Bend the piece at the scored line, separating the part from the surrounding plastic, **5**. You now have a piece surrounded by a thin black line indicating where it begins and where the sheet it was attached to ended. With a coarse sanding stick, remove the excess plastic to the black line.

Oversanding happens. The most-frequent question modelers seem to have about vacuum-formed models is "How much do I sand?" followed closely by, "What if I sand too much?" In the case of this conversion, "How much do I sand?" is answered by some of the other pieces involved. The Hasegawa nose cone, nose gear well, and windscreen show where the forward fuselage width should be, as does the vacuum-formed cockpit tub. The separate vacuum-formed refueling bump on

85

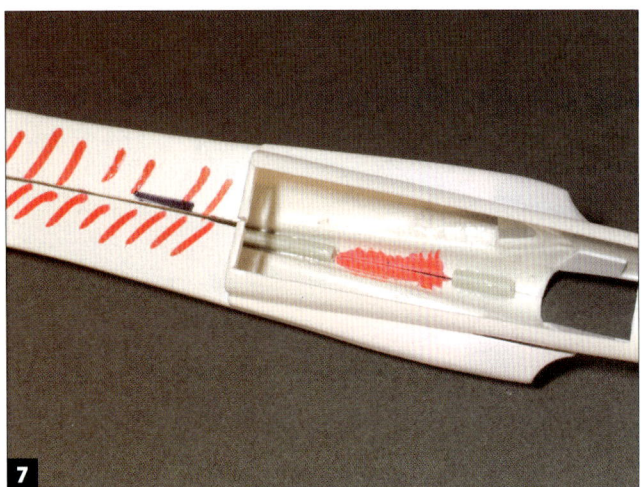

7 Here the fuselage halves are joined and a combination of super glue and sprues are used to reinforce the connection. The areas marked in red are to be removed.

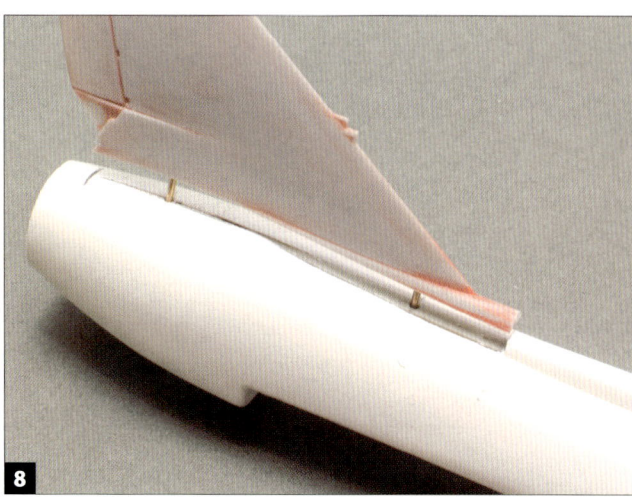

8 A pair of brass pins provide a strong, positive fit of the fin to the rear fuselage.

the spine is another gauge of width, and of course, you must consider the needs of the Hasegawa wings.

The fuselage halves should be sanded to fit these various items as tightly as possible. If one sands too much, no problem; that's what sheet styrene is for. You simply shim the fuselage halves, or whatever you're working with, back to the correct width.

Falcon provides the partial left vertical stabilizer half exactly as given in the Hasegawa kit. Trim that off; you'll eventually use the all-Hasegawa item. Only after it appears you have a good rough fit between the fuselage halves should you trim away the stock covering the cockpit opening. With this Delta Dart, the only interior supplied is a rudimentary cockpit floor and side panels. The typical approach probably would be to mount this piece to one half of the fuselage, then build as much additional cockpit on and around it as you see fit. This usually means frequent fitting of fuselage halves to ensure nothing done in the cockpit interferes with the fuselage fit. I much prefer to deal with the exterior and interior separately, but since there's no access to the inside of this Dart fuselage, that's a small problem, **6**.

Access denied? Create your own.
Remove a generous section of the belly (roughly the missile bay doors) to provide both a route to the cockpit and easy access to joints that need to be glued, **7**. Put the belly segments aside to be returned to their locations much later. I've heard it suggested to add small tabs to inner fuselage halves to help with alignment and to add gluing surfaces. This is not a bad idea, but for me, the fast-drying properties of super glue eliminate any need for this added procedure.

Starting at the nose, I carefully tack sections of the fuselage together until I get to the tail. Then I go back and reinforce joints from the inside with a combination of thick super glue, sheet stock, or sections of old sprues. With the fuselage halves so joined, you are left with a remarkably firm fuselage—every bit as solid as an injection-molded fuselage and not as brittle. And this is without any internal bracing or bulkheads. With the fuselage together and wide open in the belly, you can trim the cockpit floor to slide easily into place from behind and below.

Not surprisingly, the wings fit the vac fuselage with only the typical gap around the edges to fill. For the vertical stabilizer, I inserted a pair of brass locating pins, **8**. Where these pins will be inserted in the fuselage, there happens to be a stout piece of sprue I'd superglued behind the seam for a strong joint. Drilling two holes into that sprue gives the pins positive fitting locators.

With the fin, nose cone, and refueling bulge in place, focus your attention on filling low spots with super glue and smoothing seams and joints. In between

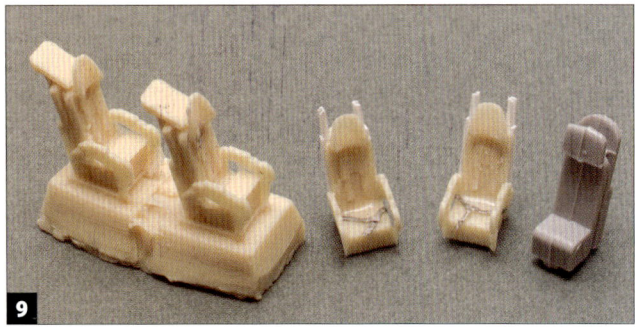

9 The TF-102 seats from left: The stock RVHP seats, the same seats trimmed and modified a little to look more like the real thing with lead foil belts added, and the Hasegawa kit seat.

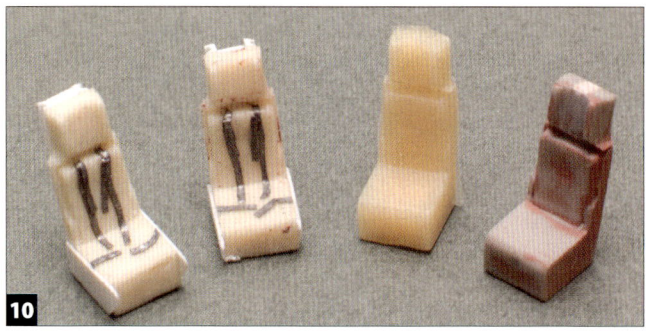

10 The F-106B seats include, from right to left, a master made from two different seats, a resin copy of the same, and a pair of resin copies with some additions to better represent the real seat.

11 On top is the section of the resin conversion fuselage the author chose to use, with the heavy rear cockpit bulkhead removed. Underneath is a stock fuselage half.

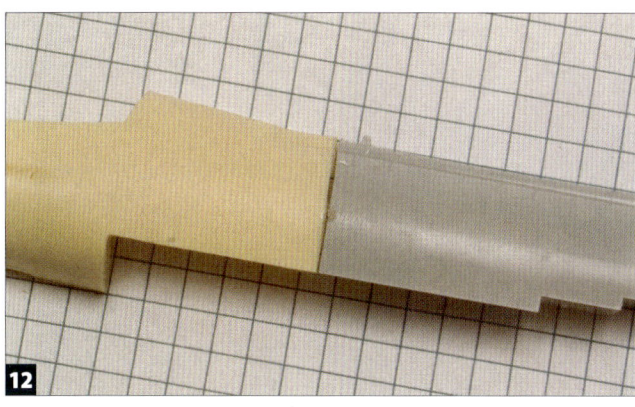

12 A sheet of graph paper is a big help when aligning a joint such as this one.

this fill-sand-prime-check procedure, you can work on the interior. For some reason, the 1/72 scale F-106 has been ignored by aftermarket-detail producers, so the cockpit and other detail for this model comes about the old-fashioned way. We'll use cockpit drawings and Monogram's 1/48 scale kit as a 3-D reference to cobble together a basic interior from strip and sheet stock, wire, and stretched sprue.

Junkyard seats. A good place to start is with seats. Hasegawa's F-106 and F-102 share a nearly identical seat, and it's not terribly accurate for either subject. The F-106 seat was unique to that type so there isn't going to be a quick substitution at hand. Using Monogram's seat as a guide, I sifted through my collection of surplus 1/72 scale jet seats for something similar in shape. I found a shape that's close enough, then modified a headrest from another seat. Bringing them together, I have a decent seat shape. Making a rubber mold from that, **9**, I made cast-resin copies and added strip styrene rails to their backs, lead foil belts on their fronts, and sheet styrene sides, **10**. They're a big improvement over the original Hasegawa equipment.

The F-106 had a prominent radarscope centered high on the instrument panel. The decking above the instrument panel reflects this with a significant bulge totally absent on the Hasegawa and Falcon kits. Carve away this area on your model, then sand a small block of RenShape to roughly match the shape you have removed. (RenShape is an expanded polyurethane foam material that has largely replaced wood in the pattern making industry. You may have to search the Web for a source, or failing that, use balsa.)

File a groove on the top of this piece and lay a piece of trimmed styrene tube in it to provide a radarscope bulge when a sheet of .015" styrene is vacuum-formed over the top of it. Once you have determined the size and shape of this vacuum-formed instrument panel hood, you can fabricate an instrument panel from sheet stock to fit the vacuum-formed piece and its place in the cockpit. Again, Monogram's kit provides great reference for what this should look like.

Referring to photos, I determined the location of the back-seater's instrument panel to put together a similar panel, less the radarscope. (That darned radarscope is in the canopy, and that will provide a challenge soon enough.) Significantly, the backside of this panel is totally exposed, meaning some attempt has to be made to represent the stuff protruding from it. I accomplished this with short sections of fine styrene rod, chopped with a single-edge razor blade on the sticky side of a piece of masking tape to keep the little bits from flying into orbit.

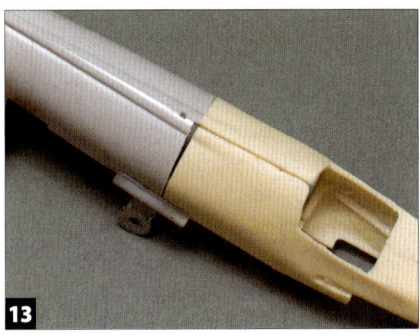

13 Aligning the resin pieces with the styrene parts results in a significant gap across the top of the resin halves. The awful gap between the resin and styrene on the right side is the result of poor craftsmanship on the part of the builder. Was he asleep at the saw, or what?

14 Fortunately, ham-fisted assembly goofs can be fixed. Styrene shims are inserted where they're needed. Note the open rear entrance to the cockpit.

15 A RenShape vacuum-forming master is fitted into the open missile bay. It came up a little short on the back end, so a styrene shim closes the gap. A thickened primer was brushed over the piece and sanded smooth to reduce the pinholes in the RenShape as much as possible.

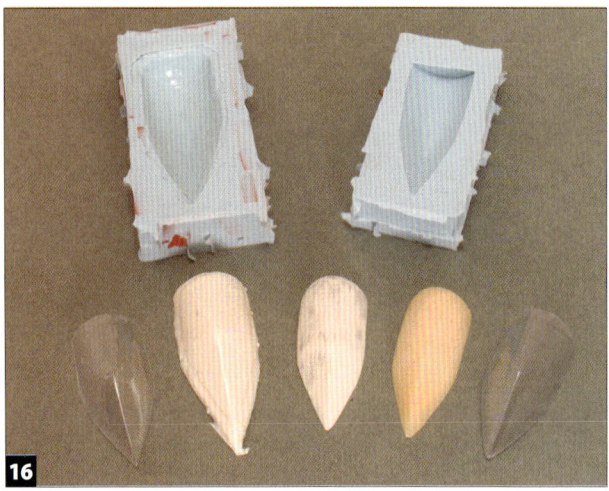

16 These items show the steps involved in making a suitable canopy. From left to right are the original kit canopy, a resin copy of the same (with the rubber mold it came from behind it), the resin copy sanded smooth and reduced in size a little bit, a new casting of the modified original (with its rubber mold behind it), and the resulting homemade canopy.

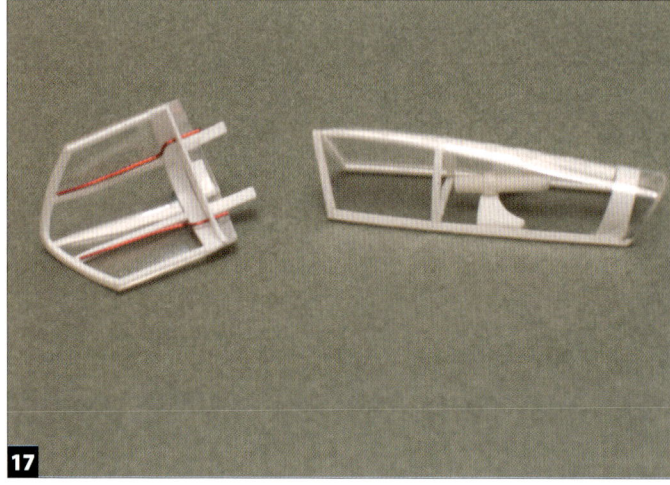

17 The clamshell portions of both canopies were modeling projects in and of themselves. Both needed internal framing from styrene strip. The TF-102 canopy also had a pair of prominent pipes running next to the top frames. Those were represented with bent wire. The F-106B canopy had a massive radarscope mounted in it for the back-seater. This was cobbled together from styrene odds and ends.

With a smear of liquid glue on the back of the instrument panel, I tweezered these tiny pieces into an orderly pattern to suggest the bodies of instruments and gauges. I also added small half-round throttle mounts on the left-side consoles. The instruments and side consoles themselves are decals. (Over time you will acquire quite a collection of instrument decals.) As the main focus of this model is a conversion demo, I didn't fret too much over precise dial placement.

TF-102

Grafting resin and plastic. A significant difference between the vac and resin conversions—other than the material they're made of—is that the TF-102 has the open missile bay as found in the Hasegawa kit. There's my access to the internals, except the cockpit, where a substantial bulkhead here must be removed. For this, I used a combination of razor saw and motor tool. With that wall gone, we can treat the fuselage and cockpit as two largely separate projects, just as with the F-106B.

Like the F-106B, the TF-102 conversion asks you to replace Hasegawa's entire fuselage. This will certainly work, but I prefer to use as much of the donor kit as possible. (I much prefer working with styrene to resin anyway.) The plan is to graft the resin forward fuselage to the Hasegawa styrene fuselage at a common panel line, exactly as I did with the Spitfire in Chapter 7 (**11**, **12**).

Again, using a combination of a razor saw and miter box, try to make four perfectly straight separation cuts. My results were mixed. The left halves are a pretty good match, so I glued these together first. Next, I added the right-side Hasegawa fuselage, which of course fits well to its left counterpart.

Now the pay-off. Attach the remaining resin nose half. The quartet of parts meeting here will reveal any problems. In my case, there were some major gaps that betrayed my lack of precision with the razor saw, **13**. I'm convinced this is one reason why Evergreen strip and sheet stock were invented, **14**. Low spots and gaps can be filled with thick super glue. Then it's time to bring the wings into the picture. Since all they have to do is click into the existing Hasegawa plastic, there should be little problem here.

RVHP supplies a pair of resin wingtips for the later Case XX (called case twenty) wing configuration if the subject Dagger is a later one, as is mine. The difference between Case X and Case XX is simply that the XX's tips drooped where the earlier ones swept up. With a pair of pliers, I slowly bent Hasegawa's upswept tips down, referring to reference photos.

Further work with sanding stick will smooth the leading edges to a constant straight line. Bite marks left by the pliers can be filled with super glue and sanded smooth. The elevons that had been removed earlier will need to be extended at a slight angle at the outboard end, so you'll need to remove a corresponding section from the trailing edge of the wing.

Your call. RVHP leaves the missile bay up to the modeler. The easiest route may be to use the Hasegawa parts and model the bay in the open position. To my eyes, these sleek delta airplanes displayed with their innards hanging out look awful. To close the bay, one can use the six separate doors Hasegawa provides, but experience has proven these doors don't come together to fill the hole very well. My solution is to fill the belly opening with a rectangular piece of RenShape, then sand and shape it to match the fuselage contours, **15**. (A piece of balsa works too.) This becomes a vacuum-forming master to make my own one-piece closed missile doors.

That's essentially the story of building this airplane. Since it's about two-thirds about building the Hasegawa kit, it was much quicker to complete than the F-106B. But then comes the canopy.

Canopies. A major focused effort on both of these models was their canopies. To be accurately reproduced, they both must be much more than clear shells. The canopy provided with the TF-102 conversion isn't

Because there are no positive fit locations for this big windscreen, I taped it in place before attempting to glue it down. Tiny amounts of medium-thick super glue were dabbed around the edges over the course of four sessions. This was followed by careful sanding to better blend the piece into the fuselage.

Since there are no aftermarket detail sets for the F-106, one has to resort to the old-fashioned method of do-it-yourself. From left to right are the main gear doors with strip styrene stiffeners added and above them a very basic blast protector for the back-seater made from brass sheet and styrene strip. Right next to it is one of two control yokes made from styrene rod, a base from shaped strip stock, and the unique Y handles of the 102 and 106 series from bent wire. Right next to the gear doors is a radarscope for the back-seater and an instrument panel mounted in a shroud that was vacuum-formed off the master to the right of it.

as impressive as the one included with the 106B. In fact, it looked like it had been made in a dust storm. The frame lines are also quite prominent, whereas on the real airplane, they were either internal or perfectly flush with the Plexiglas. Finally, with the piece not being at all protected in the box, the canopy's peak (and thinnest point) had a significant dent in it.

The first step in creating my own replacement canopy was to make a rubber mold from the original kit piece, **16**. Into the cured rubber mold I poured Alumilite casting resin. When cured, this gave me the basis for a new canopy master. Before I could use this resin piece, I needed to shrink it by at least the thickness of the clear sheet, which meant giving it a thorough and even sanding. In this process, all the raised detail was happily removed. To monitor progress, I occasionally plunked my piece into its opening in the fuselage.

When I was satisfied with the shape, size, and smoothness of my resin master, it was time to put my vacuum-forming machine to work again. It heated up a sheet of .015" K&S clear sheet butyrate, which I pulled down over the resin master. When I turn out an example in which no big dust particles or cat hairs have found their way between the clear sheet and the resin master, I've got a keeper.

To get the small rounded edges needed on some of the canopy framing, stick a piece of masking tape on a piece of sheet styrene and punch tiny holes in it with a punch-and-die set. Peel the tape dots off of the resulting styrene disks and stick them into the corners of the canopy that will need the rounded framework. Add straight sections of masking tape to essentially connect the dots (at right). Careful placement is required to get the straight and rounded edges to blend smoothly together.

Getting busy. The clamshell portion of the TF-102 canopy had a busy look to it, **17**. I added all the internal framing and bulkhead from strip and sheet stock that was ever-so-carefully superglued in place. A pair of visible pipes run parallel to the upper frames, which I reproduced with thin, easily bent copper wire.

Since I wanted to pose this canopy open, more modifications were in order.

At the rear of the clamshell section are two "ears" made from strip stock of appropriate width. These fit corresponding slots I opened in the fuselage spine. Getting the correct angles of the cuts to make these slots and the positioning of the ears in the clamshell takes fitting and refitting, with much referring to photos. When I was satisfied that the canopy looked right in the open position, I carefully blended the

21 These homemade vacuum-formed windscreens are perfectly smooth. To make their masks, the framing was sketched on the resin master for the TF-102 and traced onto masking tape, which was used to cut sheet styrene templates. Tracing those templates onto masking tape ensured (with some tweaking) identical left and right clear panels. The F-106B, though much simplified, was completed the same way.

"ears" into the top surface of the canopy with super glue, sanding stick, primer, and fine-grit wet-sanding.

Attaching the windscreen also calls for careful work. Once it had been trimmed to closely fit the upper deck of the fuselage, the gluing process began, **18**. I held the piece in place with a strip of masking tape and dabbed small amounts of medium-thick super glue all around the perimeter, attaching it and filling gaps at the same time. Using only small amounts of super glue at a time, I minimized the possibility of frosting the clear part.

Clear contrast. By contrast, the F-106B canopy is typical superb Falcon quality. I trimmed it free of its backing sheet exactly as with the P-40 canopy in Chapter 7. Like that on the TF-102, the clamshell section of the F-106B canopy was a busy affair. Since the conversion supplies only the clear part, all the business of inner framing and a radarscope and mount again comes from styrene stock. You're meant to use the windscreen from the Hasegawa kit. It works, but it won't match the vac look of the clamshell. Besides, the area directly underneath the windscreen with its new contours and height interferes with the kit piece. For these reasons, I vacuum-formed my own replacement, **21**.

With the windscreens in place, the canopy framework needed to be dealt with. Since both windscreens were perfectly smooth, I would have to create my own canopy masks. The TF-102 was the tricky one. On the resin master from which I had made the clear part, I penciled in the lines of the clear panels based on my photos. This only has to be done on one side of the master, as the opposite will be identical, only reversed.

I laid Tamiya masking tape (thin enough to be semi-transparent) over the resin piece and re-traced the lines onto the tape. I then stuck the tape to a piece of .015" sheet stock and again traced those lines with the point of a knife blade. This gave me plastic templates of my clear panels. I traced the outlines of these rigid templates on masking tape. (With these plastic templates you can cut as many identical masks as needed.) Left and right sides will be the same, or I can easily tweak the masking tape until they are. I handled the F-106B the same way, using the raised lines of the Hasegawa windscreen as my guide.

Masking and painting the clamshell portions of the canopies is a challenge. Half of what may appear to be airplane canopy frames are in fact internal seals. This fiber glass material ranges in color from tan to yellow-orange to gray depending on the airplane and, I assume, the age of the canopy. I carefully masked the clear areas of both clamshells and painted them an overall light tan. (Painting internal seals on the outside surfaces isn't exactly accurate, but in this scale, it looks acceptable.) Pulling that tape, I then re-masked to paint the metal framing light gray. No doubt about it, this is fussy work on these small pieces.

I laid down rows of rivet head decals (found on Microscale sheet 72-116, A-7Ds, or 72-121, P-51 aces) to represent the prominent fasteners seen through the Plexiglas.

Into the vortex. Then there were those vortex generators. Aarrgghh. Onto a narrow strip of lead foil, I pressed a narrow piece of strip stock. Holding it firmly in place, I carefully bent the foil against the side surfaces of the strip, removed the die, as it were, and trimmed the ends of the foil down. Having a real TF-102 at the Selfridge Air Museum to study and photograph made this process a little less daunting, but only a little. I tacked these tiny things on with white glue, **22**, not trusting myself to place each of them exactly with super glue.

Painting the airplanes was straightforward: gloss red on a base coat of gloss white for the Deuce, followed by gloss light gray on both of them, **23**. I mixed a slightly darker shade of the base ADC gray and post-shaded some panel lines on the gray areas of both, then did the same with some darker red on the red areas of the TF-102. Using this gray rather than black ensures the shading won't be very dark. For the most part, panel lines on ADC gray airplanes aren't prominent.

Decals for the TF-102 came mostly from a Hasegawa kit sheet, with the TC buzz code made from slicing up the FC provided. The small checkerboard pattern on the rudder came from a pair of 1/48 scale Hasegawa F-16 tail bands. The 57th FIS badge and ADC emblem came from old Microscale sheets. Markings on the F-106B were a mix of Hasegawa and Aerodecal, a German firm that no longer exists. I had not tried Aerodecal's products until this project, and found the decals to be quite stiff and brittle. The decals only conformed to the drop tanks with large amounts of Solvaset and a couple tiny dabs of super glue as well. The red and white rudder stripes turned out to be significantly oversized, and I ended up doing

22 This close-up photo of the TF-102's front end shows the row of tiny vortex generators that have been cut from lead strip and attached with white glue to its raised canopy. The finished plane also reveals how much easier it is finish an item like its "barber-pole" pitot tube off the plane and install it carefully to the nearly complete model.

23 The planes are now ready for decals. Both models have been lightly post-shaded with a slightly darker shade of the base ADC gray mix to suggest some panel lines. The radomes are unpainted, waiting to be dealt with at least until the intensive handling involved in the decaling process is completed. Decals will be a mix of kit and aftermarket items. Some, such as the T in TC (page 93), will be creatively pieced for the correct "buzz numbers."

them with white paint and strips of red decal stock.

The landing gear on both the Dagger and the Dart featured retraction arms and struts to various doors not provided in the Hasegawa kits. These were added from stretched sprue and strip stock. Also added were landing lights inside the main gear door (from shaped styrene rod) and a tiny disc of chrome tape punched out with my punch-and-die set. Once again, the Monogram 1/48th versions of both jets provided good reference for these landing-gear details.

Removable pitot. Only with the bulk of the decals in place and most of the landing gear attached did I deal with the radomes and pitot tubes. Drilling a hole into the nose of a model can render it quite fragile and prone to chipping if it bumps into something, and it's best to not risk weakening the front end this way until nearly the end of the project. The process is simple enough: Center and drill a hole to the approximate size of the tube you plan to install, then ream it to the exact diameter, **24**. I recommend placing—but not gluing—items like these, **25**. This allows you to remove the item for handling the model.

The canopy clamshells are likewise set in place without glue, **26**. A flaw in the Falcon F-106B canopy then revealed itself. The real airplane had a red anti-collision beacon on the back of the canopy that was visible when the canopy was open. I put the light on mine, but at the correct open angle, it's hidden inside the fuselage. I placed mine in relation to the rear windows as I saw in photos, and now it appears the windows extend too far back. Oh well. I'll make adjustments on the next one.

24 After a precisely centered pilot hole is drilled into the tip of the radome, a reamer is used to enlarge it to precisely the diameter of the pitot tube to come.

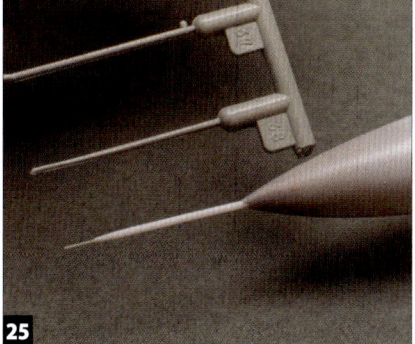

25 The pitot tube is a length of styrene rod. One end is carefully drilled out, then tapered down some with sandpaper. Inserted into the hole is a short piece of stretched sprue. Here it's test-fitted into the nose, with the Hasegawa kit rendition still on the sprue for comparison.

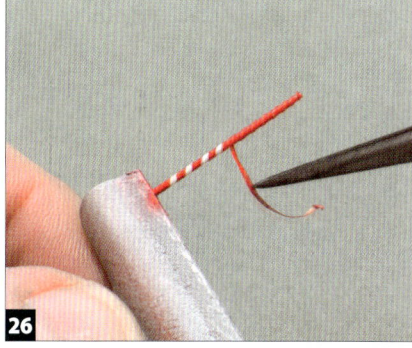

26 To get the classic barber-pole effect on the pitot tube, a thin strip of very flexible masking tape is evenly wound around the white rod. After a coat of red paint, the tape is unwound and voilà! This is much easier than those spiraled Luftwaffe spinners.

0-72537
U.S. AIR FORCE

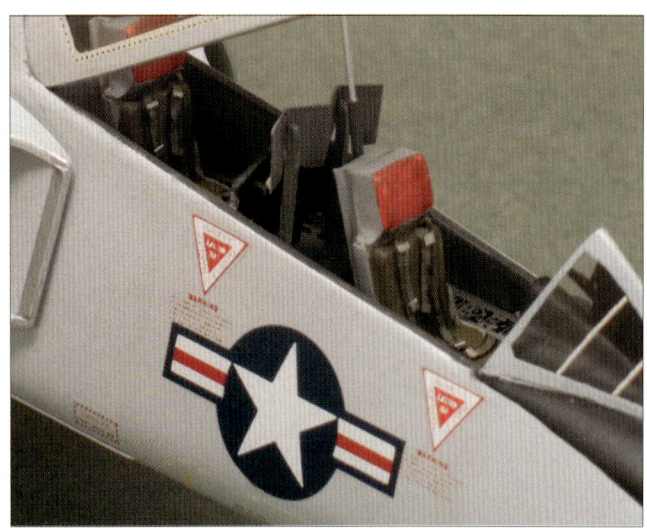

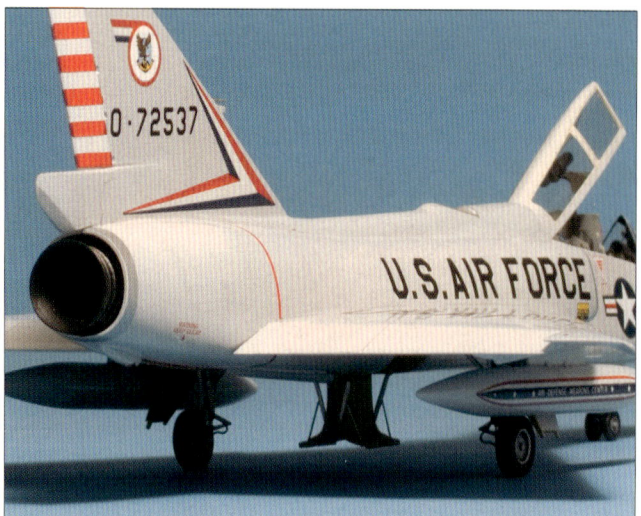

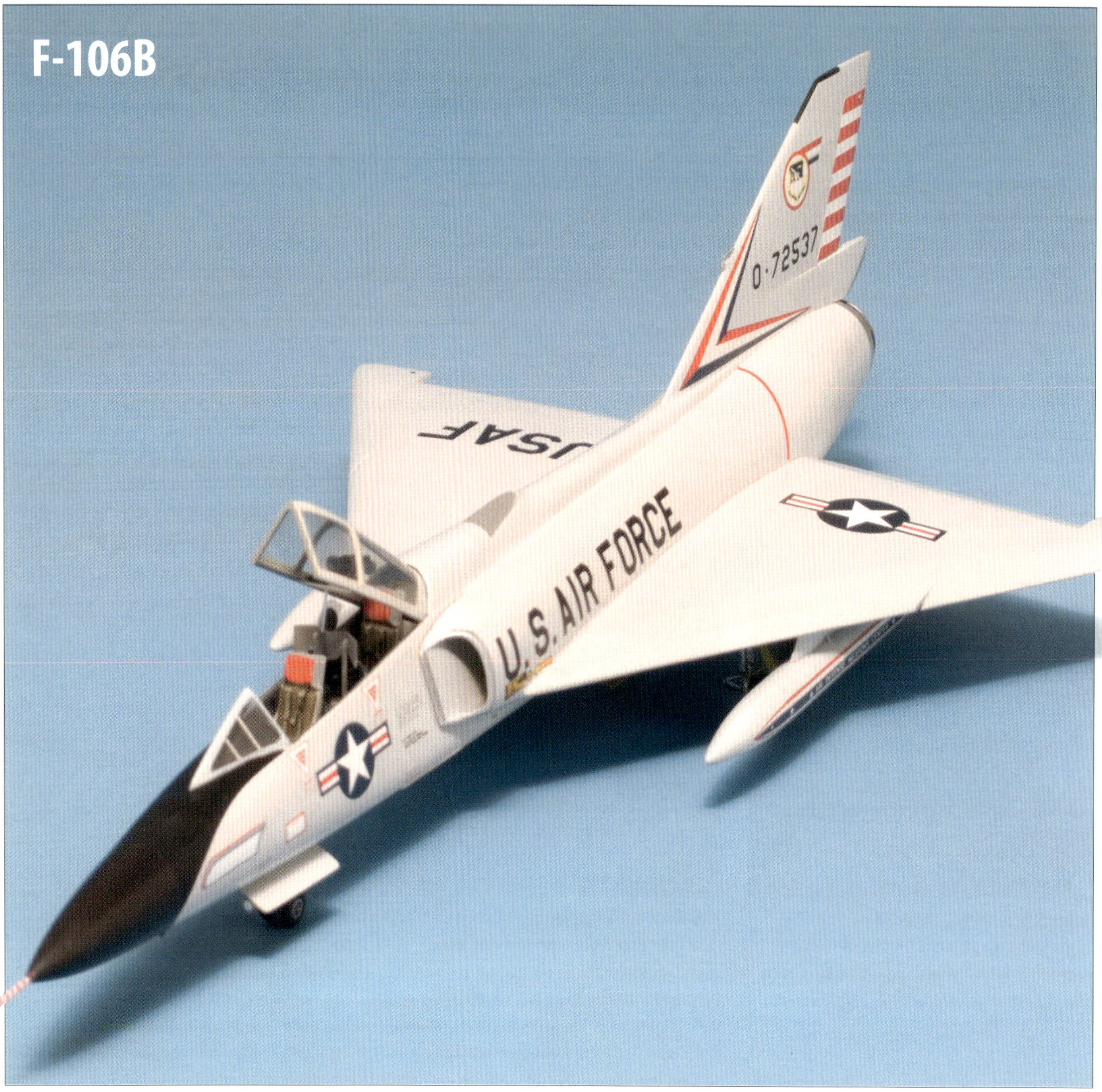
F-106B
0-72537
USAF
U.S. AIR FORCE

TF-102

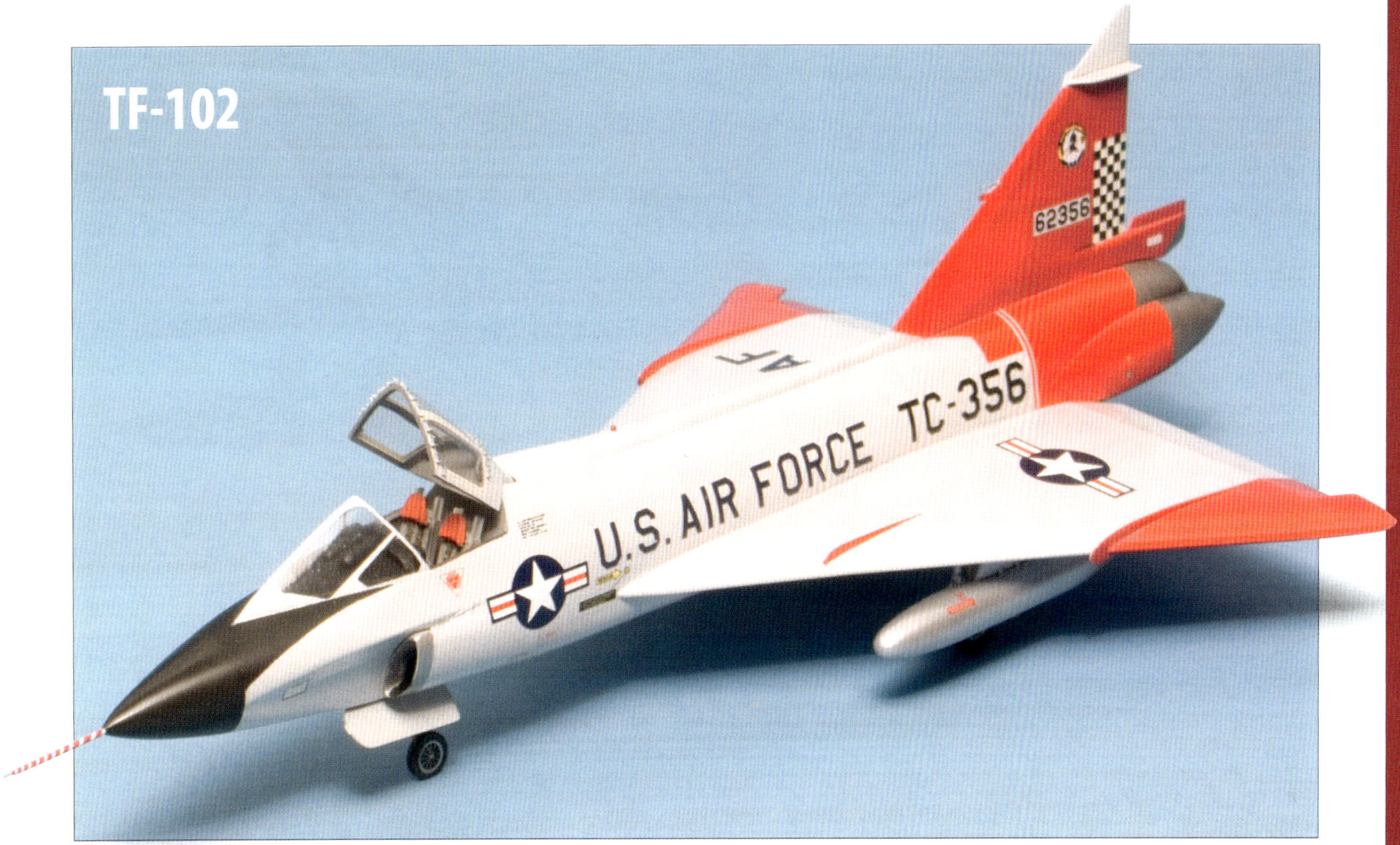

62356
AF
TC-356
U.S. AIR FORCE

U.S
RESCUE

62356
TC-356
U.S. AIR FORCE

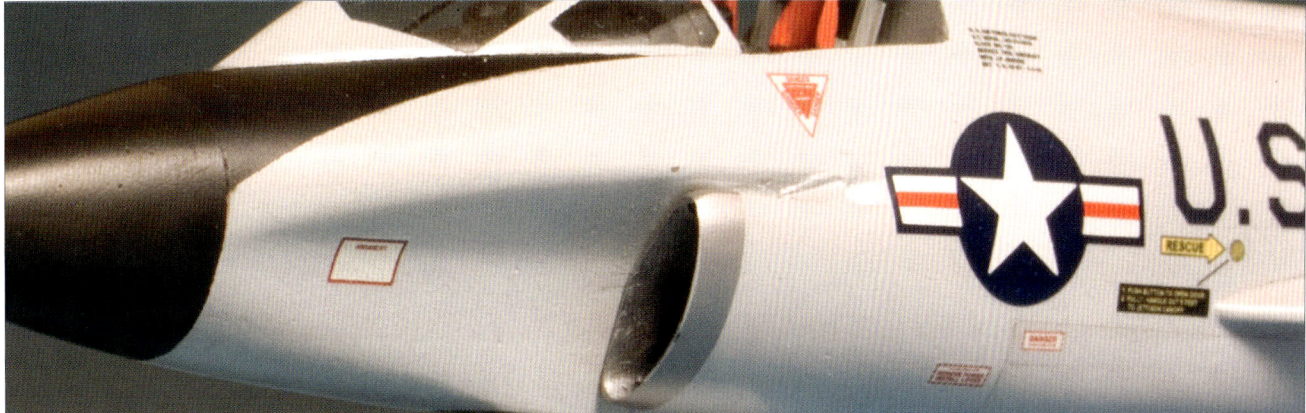

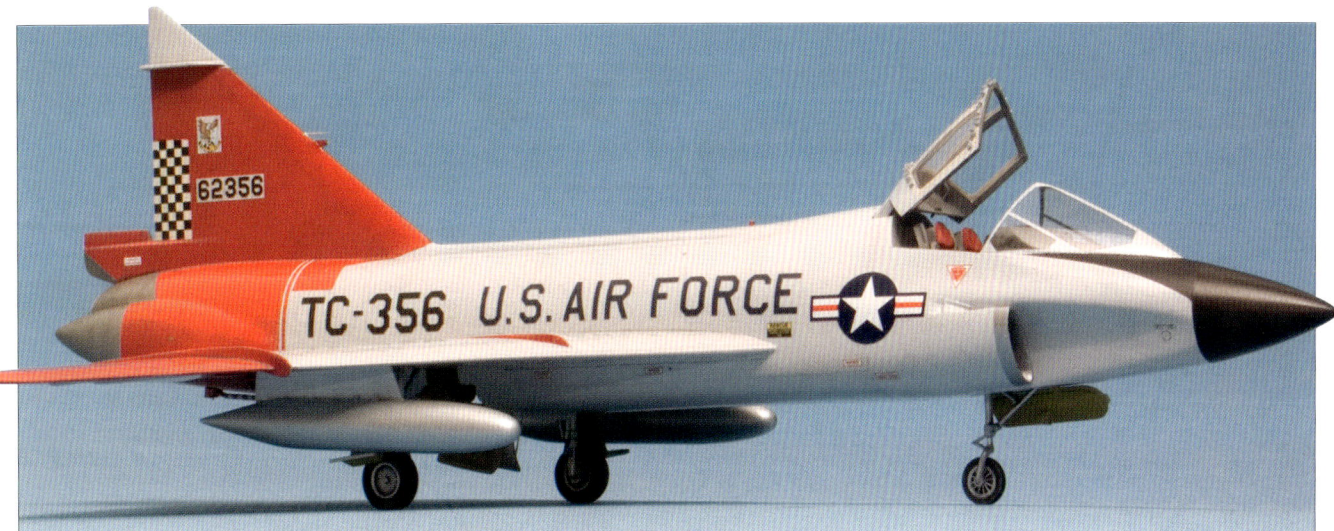

Monogram 1/48 scale F-100F "Hun"

Hasegawa 1/32 scale Bf109E

Monogram 1/48 scale F-5E Lightning

Dedication:
To Molly Hunt, wonder wife and primary editor.

About the author
Pat Hawkey is a child of the '60s, back when plastic models were entertainment for most young boys (and even a few girls). He remembers models being found everywhere toys were sold—at hardware and drug stores, groceries, a local camera store, and of course, hobby shops.

Pat was raised only a few miles from the Detroit suburb of Hazel Park, Michigan, where the Squadron military hobby empire began. Besides its kits and books and accessories, the Squadron Shop featured huge (especially to a young boy), beautiful glass cases full of finished model planes, tanks, and figures. His goal became to make one of his models good enough to be displayed in those cases too. That happened when he was still in his teens, then one day in his early 20s, the store manager told him somebody was looking to buy a couple of his models. Ever since then—in one fashion or another—Pat has been a professional model builder.

At age 30, he started Hawx Planes, hoping to make his modeling into a career. Fortunately several serious collectors came along who had long wish lists. One of them entered one of his models under Pat's name in a modeling contest. It won an award and was photographed by and published in *FineScale Modeler* magazine. That led to his first model construction article being published in the magazine in the mid-'80s. Today, Pat is a regular contributor to and advisory board member for the magazine.

Pat and his wife and a couple of rescued dogs and cats live in Temperance, Michigan.

At left. The author says these models (all of which have been featured in *FineScale Modeler* magazine) show that with simple tools, basic modeling skills, an eye for aircraft shape, and the application of the appropriate level of detail, anyone can create realistic scale model aircraft worthy of display.

It's all in the details –
Build, paint, and finish your models to perfection!

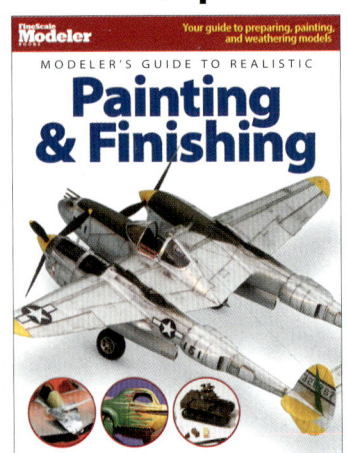

Perfect for the beginning plastic modeler who wants to learn more advanced techniques, this photo-driven guide includes an introduction to airbrushing and sections on brush painting, spray painting, weathering, and applying finishes. 80 pages.
12257 • $16.95

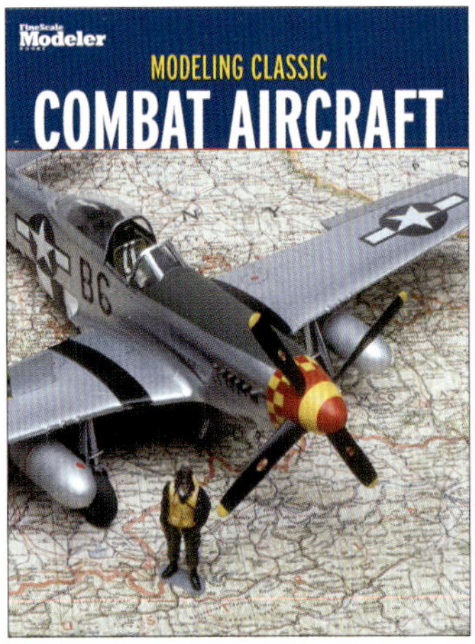

Easy-to-follow diagrams, crisp photos, and concise instructions teach you to build, paint, and finish military aircraft models from WWII to today. The more than 15 step-by-step projects include the P-51 Mustang, F-18 Hornet, the A-7D Corsair, the B-17 Flying Fortress, and more. 96 pages.
12255 • $21.95

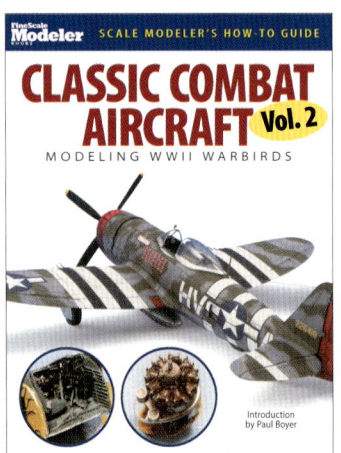

This second volume features 18 projects that include Gabby's Last Jug, a D-Day Spitfire, an American Mosquito, and a late-war Messerschmitt Bf 109. Build, modify, detail, paint, and weather legendary aircraft with help from the *FineScale Modeler* experts. 96 pages.
12431 • $18.95

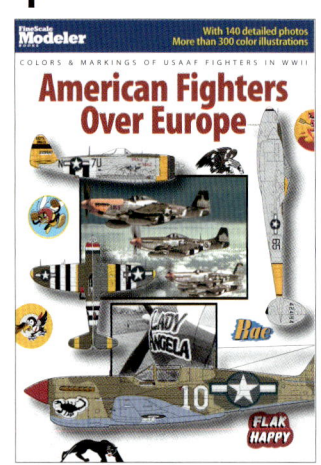

This guide describes colors and markings for U.S. Army Air Force fighters in the European and Mediterranean theaters of WWII. With more than 300 illustrations and 140 photographs, the book makes an excellent reference for modelers and aviation buffs alike. 112 pages.
12427 • $21.95

Buy now from hobby shops! To find a store near you, visit www.HobbyRetailer.com

www.KalmbachBooks.com or call 1-800-533-6644

Monday – Friday, 8:30 a.m. – 4:30 p.m. CST. Outside the United States and Canada call 262-796-8776, ext. 661.